THE GENTLE
SEA

THE GENTLE SEA

DEE SCARR

Published by PADI, 1251 East Dyer Road #100 • Santa Ana, CA 92705-5605

Library of Congress Card Number 90-060543
ISBN 1-878663-00-3

Printed in the United States of America
10 9 8 7 6 5 4 3 2 1
PRODUCT NO. 70140

The Gentle Sea is dedicated to my best friend, David (even if he does think "phylum" is something you do with papers you don't want lying around on your desk . . . or maybe **because** he thinks "phylum" is something you do with papers you don't want lying around on your desk . . .).

Acknowledgments

The following friends have been more than generous with their time and knowledge, and were sympathetic despite my occasionally irrational behaviors as *The Gentle Sea* was being written. My thanks to Linda Goodwin, Terry Kardos, and Andrea Struss, who cheerfully read the (very) rough manuscript, and to Hillary Roberts and Genie Clark, who read the smooth one; to Stephen Lewis, George "Clam" Buckley, and Roberto Hensen, for listening to my lunatic-fringe theories; to Nancy Laughlin for the wonderful image of squid as hummingbirds; to Diane Batalsky, Howard Batalsky, Jon Fein, Enid Lewis, Pat Mason, my parents, Rose and Carl Scarr, and Belle and Herman Rosen, for believing; to PADI and particularly CK Stewart, Barbara Mourant, and Karl Shreeves — for giving me a deadline!; to all the people of Bonaire, for protecting the marine environment in which I have learned so much; to Crooked Jaw and Ms. Piggy and the Cookie Monsters for accepting my friendship; and to each and every *Touch The Sea* diver.

Table of Contents

Introduction

For years, the impression of the undersea world given to the general public — including entry-level scuba students — was that of an environment filled with creatures not just capable of, but interested in, biting, stinging, and otherwise injuring humans. The programs on television have done little to dispel this image, and my conversations with underwater cinematographers have confirmed that the media is interested mostly in films of animals that are big or dangerous, and preferably both.

Yet, from the point of view of those of us who spend a lot of time under water, this view is essentially false.

That is, the sea is *not* filled with animals hoping to hurt humans who have entered their domain. The sea is filled with animals minding their own business: foraging for food, seeking out mates and mating, avoiding predation, and so on. (Norine Rouse tells her beginning scuba students: "People are in the sea by virtue of technology, not biology; therefore, we are neither the natural enemy nor the natural food source of any creature in the sea.")

The everyday activities of marine animals fascinate me, and the more I understand of these behaviors, the more interesting diving — and snorkeling — become. For example, watching a bunch of little fish swimming up and down in the water column might keep my interest for only a few minutes, but understanding what they are doing can keep my interest for hours. Once I understood the harem structures of razorfish and their basic spawning behaviors, two hours at 15 feet watching razorfish wasn't long enough!

When anyone first enters the sea with scuba, simply seeing all the critters in their natural environment is fascinating. As the sensory overload lessens with more dives, divers begin identifying what they are seeing. They learn which animals are common and which are uncommon or rare. (One of the most frustrating experiences a divemaster may have is seeing an unusual animal on a dive with novices; when the group surfaces the divemaster enthuses over the sight of a rare eel, for example, or an unusual anemone, while the new divers exclaim over the [extremely common] arrow crab the divemaster showed them.)

After divers have learned the names of the obvious creatures, they begin to search for hidden animals such as the shrimp that live in anemones, or crabs and lobsters under ledges. After that they generally begin to discover the camouflaged animals: crabs that attach sponges to their shells to match the sponges they live upon, rock-like scorpionfish, nudibranchs that look like little sponges.

It is only after they have mastered the finding of animals that most divers begin to pay attention to the behaviors of those animals, and because the behaviors can be puzzling, many divers become bored at this point in their underwater activities.

The goal of this book is to help you, as a diver, snorkeler, and/or naturalist, understand more clearly the marine environment and the behaviors of the animals who inhabit it. If you're thinking that it would be impossible to discuss all the behaviors of all the animals, you're correct — of course. But in this case a little information isn't a dangerous thing; rather, it can help you to go beyond understanding specific behaviors to figuring out other behaviors that you may see. In a sense, we can all be underwater detectives: What is that animal doing? Why? What will the result be? How does it fit into the total picture?

If this book is successful, you'll never be bored under water.

Preface: Organization

One of the most difficult challenges in explaining the behaviors of marine animals is how to organize the subject matter. We can't understand behaviors until we can identify animals; but even then, there are the factors of time of day, the various habitats, and how animals interact with other animals — both of their own and other species.

I learned about marine animal behaviors not by compartmentalizing, but by spending hours in the water watching the animals and trying to figure out what I was seeing, and by spending hours in the water actually interacting with various marine animals to learn what they would do under various circumstances. When I'm not diving, I'm often poring over books about marine animals (an annotated bibliography of some of my favorite books begins on page 144), discussing observations of my own and other divers, or picking the brains of marine scientists.

To present this material in a way that will make any sense, organization is necessary — but overlaps will occur frequently; don't expect a dry text from someone who spends so much of her life immersed in water!

The Gentle Sea begins with a discussion of the phyla (singular = "phylum" — the major divisions of the animal kingdom) whose members are of particular interest to divers. We'll examine the basic life-support systems of each group: how they eat, how they avoid predation and how they procreate.

We'll seek out some particularly interesting interactions between marine animals who are dependent on other marine animals for their lives — commensal relationships such as the one between the clownfish and the anemone in the Pacific oceans.

We'll consider how the times of day affect the behaviors of the animals: dawn, daylight, dusk, and dark.

Finally, we'll address the part of diving that concerns me most: how to be a gentle diver.

The animals specifically discussed in *The Gentle Sea* are those that inhabit coral reefs, particularly the reefs of the Caribbean, and especially those of Bonaire (where I have done more than half my life's diving). Nevertheless, once you understand the behaviors and interrelationships of one set of coral reef animals, you can easily extrapolate the information so that it applies to the animals that inhabit the waters anywhere.

SPONGES
(Phylum Porifera)

Imagine my surprise when the tube sponge spit out the film can! The film can was just the right size to fit into the end of a purple tube sponge (the lid was wide enough to keep it from falling down into the tube), and it seemed like the perfect place to hide a prize for our underwater film can hunt — until the sponge refused to cooperate. I put the film can into the end of the tube again, and the results were the same: it popped back out. Finally I gave up and hid the can elsewhere. I'd never paid much attention to sponges except to be aware that they made nice photo subjects. If I'd known then what I know now, the spitting sponge behavior would have made more sense.

A diver examines a large tube sponge off Little Cayman.

The Phylum *Porifera* ("pore-bearers") consists solely of the sponges, which, in terms of evolution, are a dead end; that is, nothing further evolved from sponges. Despite the fact that sponges can't move around — as a matter of fact, adult sponges must be attached to live — sponges are animals.

Sponges are perfectly adapted to do what they do, which is probably why they haven't changed in about 200 million years. There are about 5000 species of sponges, mostly marine, and they live in shallow as well as deep waters; one of Dr. Eugenie Clark's favorite sights off Grand Cayman — 3000 feet deep via submarine — is a huge white sponge shaped like a flower.

Sponges have no true tissues or organs; rather, each sponge consists of three types of cells: an outer layer of covering cells; an inner mass of flagellated cells, which pull water through the sponge and capture and ingest food; and a group of amoeba-like cells, which move food particles. Some sponges keep their shapes with an internal skeleton of glasslike spicules, others have only stringy threads of a protein called spongin, still others have both spongin and spicules. All sponges have lots of holes — the canals through which water passes.

Sponges are nature's ultimate water filters, taking the tiniest planktonic critters from the water for food and spewing the filtered water back into the sea. An average-sized sponge filters hundreds of gallons of water in a day.

The manner in which sponges filter water leaves them vulnerable

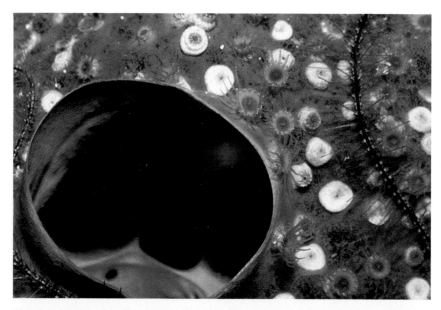

This 1:1 closeup of a Little Cayman sponge shows some of the sponge's interior structure, its commensal zooanthids, and parts of the arms of a resident brittlestar.

to problems from silting: once silt (such as that stirred up by a storm or kicked up by divers) covers their intake ports, they have no way of removing it, and the passage of water — and nutrients — into the sponge is blocked. Some of the silt falls off of its own weight; some is kicked off accidentally or dusted off deliberately by divers; but if too much silt remains the sponge will die of starvation.

Some sponges have only small exhaust ports all over their structures, but others, such as barrel sponges and tube sponges, combine many small exhaust ports into one large one through which their filtered water passes with surprising force — as I discovered on my treasure hunt preparation.

Sponges can reproduce sexually or asexually. In sexual reproduction, the male sponge releases large numbers of sperm into the water; the

I caught this purple tube sponge "smoking" (ejecting gametes into the water) on Ebo's Reef, off Klein Bonaire.

sperm enters female sponges and fertilizes their eggs. I first noticed this phenomenon when swimming along a section of reef that had much lower visibility than the rest of the reef. As I looked around I saw many sponges, all of the same type, "smoking." I had a feeling there was an orgy going on, and I was right.

Although some sponges can produce both sperm and eggs, they generally do so at different times so that cross-fertilization, instead of self-fertilization, takes place. Eventually the sponge larvae are released into the water, and those that land on a hospitable spot develop into sponges; the rest die.

If a piece of sponge is broken off and lands on an appropriate place with little water movement, it will attach itself to the bottom and begin a separate formation.

I found my favorite illustration of the talents of sponges in Kaplan's *A Field Guide to the Coral Reefs* (one of the Peterson Field Guide Series): In the early 1900s a sponge was separated into its individual cells and left in a container (presumably with sea water). In only one day the cells had managed to gather together, and eventually the sponge rebuilt itself to almost an exact replica of the original. (Can you imagine the results if Hollywood learned about this? *The Sponge that Wouldn't Die,* a new film by. . .)

Sponges come in about every color imaginable, from bright red to brown to lavender to green to yellow. Sponges of the same species can grow in different shapes, depending on prevailing currents and other growing conditions. Thus, even biologists have trouble determining exactly what kind of sponge they're looking at (unless they're looking at it under a microscope). Common names for sponges are uncomplicated: the first adjective is usually the color, the second adjective is the general

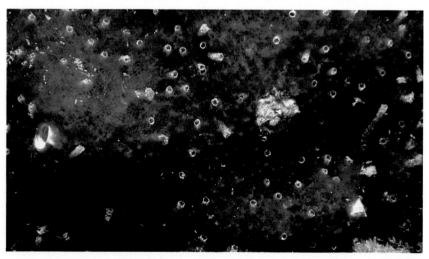

The do-not-touch-me sponge in close-up (1:2).

shape, the last word is the noun *sponge.* Thus, we have purple tube sponges, orange encrusting sponges, green finger sponges, and so on.

Another type of common name for sponges is of particular interest to divers: the warning name. In the Caribbean there are two types of sponges which can cause pain and blistering to the bare skin of marine *Homo sapiens:* the fire sponge (*Tedania ignis*) and the do-not-touch-me sponge (*Neofibularia nolitangere*). Each of these sponges is red, by the way — the fire sponge is bright red and lives in the calm waters of bays and lagoons, and the do-not-touch-me sponge is brick red and stocky, and lives around coral reefs.

An interesting characteristic of sponges is that they are not eaten by most other reef inhabitants. Filefish and angelfish are said to be the major sponge predators; in my experience even these fish prefer eating sponges that are damaged (by an anchor chain or a diver's fins, for example). There must be something in the outer layer of sponges that's unpalatable; once that outer layer is broken, fish are much more interested in eating sponges.

Sponges can be large or small, live everywhere on the reef, come in virtually any color, provide a huge amount of additional surface area on a Caribbean reef, and are generally ignored as a food source. Thus they provide an important ecological niche utilized by many other reef animals. Many small fish and invertebrates spend their entire lives within the cavities of tube and vase sponges, as do some zooanthids (coral-like animals). Brittle starfish spend their days within the cavities of sponges, and their nights on the outside of those sponges filtering food from the water.

Other animals use living sponges for camouflage, or camouflage themselves to look like sponges instead of the tasty morsels they really are. The more I learn about the ecology of the reef, the more sponge influence I discover.

The underside of this ledge off Little Cayman supports a great variety of sponge growth.

STINGERS
(Phylum Coelenterata)
The corals, anemones, and jellyfish

Think about the criteria most of us use, consciously or not, to define an animal: 1) we expect it to be able to **move**, and 2) we expect to be able to see it **eat**.

Since the life-supporting processes of corals and (to some extent) anemones are subtle — though not so subtle to us as that of the sponges — to many divers the reef-building corals seem to be rocks and the anemones plants. By understanding them better, though, we'll be able to see

I believe many Caribbean fish are "investigating" the possibility of living with anemones as anemonefish do in the Indo-Pacific. This is a giant Caribbean anemone.

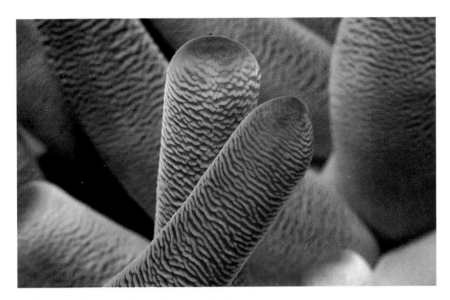

A closeup of the tentacles of the giant Caribbean anemone.

them as animals — and more like animals (as we define them) than sponges, because they're more advanced.

Anemones

Anemones are very adaptable. They live in the warm waters of the tropics as well as the cold waters off California and New England; they live on coral reefs, rocky bottoms, and in sand flats; they come in colors ranging from almost white to brown to red to purple; they have tentacles in varying lengths and arrangements around their mouths.

Feed an anemone a piece of hot dog. If you merely hold the hot dog in front of the anemone, it can't sense the food and doesn't reach for it. But, if you touch a tentacle with the hot dog, the tentacle will stick to the hot dog, and, by contracting, will draw the hot dog into the anemone's body cavity. If the food you offered was a whole hot dog, at first part of the hot dog would end up sticking out of the mouth of the anemone, and if you looked carefully at the stalk of the anemone, you could see the rest of the hot dog inside.

Anemones display the characteristics of their phylum, *Coelenterata,* which literally means "hollow gut." They, like all other coelenterates, are radially symmetrical; they have a single body cavity that both digests food and circulates nutrients; they have a single orifice through which food goes in and waste material comes out; they have stinging cells. Most anemones have a stalk with one end attached to the substrate and the other, the mouth/anus end, surrounded by a ring or rings of tentacles.

The coelenterates' stinging cells, specialized for feeding and defense, are called *cnidocytes* (some biologists call this phylum the *Cnidaria* rather than *Coelenterata*); within each cnidocyte is a capsule called a *nematocyst*. When the cell is activated (by touch, or chemical stimulus, or even nerve impulses of the animal itself), pressure inside the capsule causes the nematocyst to discharge — like the fingers of a rubber glove popping out when you blow into it. The barbed harpoon of the nematocyst penetrates the target and envenomates it. Each nematocyst can only fire once, and it may take hundreds of them to have any effect on the coelenterate's prey.

However, the stingers of a particular anemone or coral or jellyfish are usually enough to capture its prey. Once the critter is captured, the tentacles of the coelenterate bring the food into its mouth.

Nematocysts also protect the anemones and their relatives. If a bunch of little fish become interested in the hot dog you're feeding to an anemone, they'll try to pull the food out of the anemone — without touching the tentacles. As the anemone draws the food further inside itself — as more and more tentacles make contact with the food — the fish stop trying to steal it because the food isn't worth the stinging they get from the anemone's nematocysts.

One of my favorite types of anemones is *Stoichactis helianthus,* called the "sun anemone" by the world (since that's a partial translation of its scientific name), and the "sticky anemone" by me (because its tentacles are stickier than most anemones' tentacles). The increased stickiness of the sticky anemone's tentacles means that their nematocysts are more potent than those of most other anemones. Another difference between this anemone and most others is that its tentacles are very short; a tentacle

Anemones will attempt to eat almost anything organic, but I was surprised to see this one try to eat a toy plastic fish.

can't move food into the sticky anemone's mouth because the tentacles are too short to reach the mouth! So, when a sticky anemone gets a piece of hot dog, the anemone simply folds around the food. It looks as though the hot dog is sinking into quicksand.

Wrasses and gobies try to steal food from the sticky anemones, but they are much more cautious than they are with other anemones. I saw why they needed to be so careful on the day that a two-inch goby got too close — instantly the sticky anemone's tentacles held the goby captive. The goby's struggles enabled more tentacles to contact it, and it soon disappeared into the anemone.

The forked anemone (also called the bifurcated or branched anemone, *Lebrunia danae*) is unusual in that it has two different types of tentacles: one type, which extends at night, catches planktonic food; the other, which extends in the daytime, contains symbiotic algae. Like the sticky anemone, the forked anemone has more potent nematocysts than most Caribbean anemones. Also like the sticky anemone, it's fun to watch: as the "changing of the guard" takes place between the two types of tentacles, this anemone seems to blossom.

On just one reef (Lamachaca Reef) on Bonaire, there are giant Caribbean anemones, orange-ball anemones, tube anemones, forked anemones, white-spot anemones, and ringed anemones. There is also a cluster of anemones that I've never seen anywhere else: they are basically cup-shaped, greenish in color, and instead of long numerous tentacles have only a few nubs on the edge of the cup. I realized they were anemones

This large orange-ball anemone lives on a piling beneath Bonaire's Old Pier.

only after touching the inside of the cup lightly; the outer edges drew in to imprison what might have been prey (but was actually my touch), and showed me what kind of critter it was. Later I was able to find it in my trusty Colin; it is *Paradiscosoma neglecta,* and Venus flytrap anemone seems appropriate as a common name for it.

While diving in the Coral Sea I found something I couldn't identify: a smooth, round ball attached to a rock. At the end of the ball, facing away from the rock, was an opening, and the lip of the ball folded back upon this opening. I ran my fingers beneath the lip and found two shrimp hidden there! When I removed my hand, one of the shrimp grabbed the edge of the lip with its claws and pulled it back over itself like a child rolling up in a blanket!

The next day I returned to the spot, camera in hand, to photograph this phenomenon for posterity — but there was no ball! There was, however, a flat anemone with short, sparse tentacles. I felt around the edge of the anemone for the shrimp, and caused the anemone to close up a bit. It began to look a little like the ball I'd been looking for — yet no amount of disturbance would cause it to roll up as it had the day before. With everyone on the *Coralita* alerted to look for the mysterious football anemone, we finally found a few more. It seems that this anemone lies flat during the day and purses its edges together with the coming of night. The mystery is not entirely solved, since the anemone isn't shown in any of the books we had on Coral Sea marine life, but at least we were able to figure out what kind of animal it is.

Most anemones keep their forms twenty-four hours a day, but, like the football anemone, a couple of Caribbean anemones change their behaviors with the time of day. Cerianthid (tube) anemones appear on sand bottoms at dusk as if by magic, but they actually extend into the water from their tubes in the sand. Tube anemones range in color from milky white to brown but they all have two rows of tentacles: an inner ring of short tentacles surrounding the mouth, and an outer ring of longer tentacles.

The outer tentacles seem to be responsible for sensing danger, and capturing food and delivering it to the inner tentacles. A gently touched outer tentacle will whip inward to carry the food (it "thought" it caught) to the inner tentacles. After varying numbers of these false deliveries (my personal record is eight) the anemone seems to realize it's being fooled — and retracts suddenly into its tube.

The tube anemone's inner tentacles are much less sensitive; they'll attach to a diver's finger and accept quite a bit of downward pressure before reacting.

Cerianthid anemones in direct light demonstrate their preference for darkness: their normally straight outer tentacles curl more and more until they resemble a collection of corkscrews, and then — the anemone disappears beneath the sand.

The orange ball anemone (*Pseudocorynactis caribbeorum*) isn't

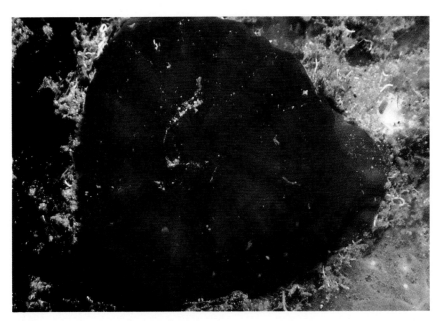

The orange-ball anemone is a nocturnal feeder; in the daylight (or after being disturbed) it pulls its tentacles inside itself.

much easier to find in the daytime than a tube anemone — it looks like a small brown lump. At night the brown lump unfolds into a brown-stalked anemone-like polyp with transparent tentacles. At the tip of each tentacle is (as you might have guessed) a bright orange ball about the size of a roller skate ball-bearing.

The color of the orange ball anemone makes it a favorite for photographers, who soon learn how light-sensitive it is: after one shot or at most two, the tentacles fold gracefully inward like a budding flower in reverse — and only the brown lump is left.

Corals

The standard anemone body design is called a polyp; stony corals and gorgonians and soft corals — the anemone's relatives — have the same body design. Anemones are solitary animals, though, while most corals are colonial. One coral formation may consist of hundreds of polyps all living together.

The coral reefs that are characteristic of tropical diving need warm waters in which to grow; generally they are only found within twenty-five degrees of the equator. Stony corals secrete a calcareous limestone skeleton, and most stony coral polyps retreat into their limestone cups in the daytime and expand to feed only at night. When thousands of individuals in a hard coral formation expand their polyps, they create an efficient

◄ *A single orange cup coral polyp, 1:2.*

A diver swims over two unusually large mounds of clubby finger coral.

"net" for capturing the plankton they eat.

If you'd like to see stony corals at work, the time to do it is at night. You need a bright light; if you hold it still for a couple of minutes it will usually attract lots of planktonic critters. Once you've attracted the plankton, slowly move the light near a coral head with expanded polyps, and you'll see the coral tentacles grabbing up the plankton just like miniature anemone tentacles.

Feeding brain coral.

A Bonairean sea fan.

Since coral polyps are generally retracted in the daytime, however, an awful lot of divers think of coral heads as rocks — to the detriment of the coral. When coral polyps are retracted they're protected from coral-eaters like butterflyfish, but not from divers. Scuba instructor David Serlin makes the point: "You know how, when you touch the coral, slime comes out? That slime is **coral guts**!" The pressure of a diver's weight cuts the coral polyp against the edge of its own skeleton. Also, slime is secreted by the coral in an effort to remove what is damaging it. Once a few of the polyps in a coral head are killed, algae can begin to grow in their places,

A deepwater gorgonian, a type of horny coral, grows along the dropoff of Bonaire's Playa Benge.

and the entire coral head is endangered.

No divers deliberately damage corals. Their difficulty is in learning the difference between coral and rock. One useful guideline is: until you're sure what you're touching, don't touch anything — with your hands, with your fins, with your knees, or anything else!

Once you learn a bit about corals, they're easy to identify. Living stony coral formations have clean-cut edges and are very convoluted.

Beneath a ledge in the Coral Sea, David Batalsky photographs sea fan growth. A crinoid is attached to the fan in the foreground.

◄ *Variously colored soft corals cover this section of Red Sea reef.*

Usually the living coral has some color (provided by a symbiotic algae called zooxanthellae), such as the browns and greens in the Caribbean and the blues and pinks in the Pacific. You can examine photographs of living corals or find some that are expanded (feeding) at night. Wave water gently over the coral head and the polyps will retract, returning the coral head to its daytime appearance. Probably the best method of learning to recognize coral, though, is to recruit someone to help you.

What we in the Caribbean generally call *soft corals* — the sea fans, sea whips, and gorgonians — are technically *horny corals*, named for the material that gives them their texture. Like the anemones, these are often perceived by divers, with some justification, as plants: gorgonians are attached to the bottom, wave back and forth with the current, and appear to have branches. A close look will reveal that each of the branches of an undisturbed gorgonian is covered with tiny polyps. If you gently touch one part of the branch, the polyps will retract in both directions from your touch; if you touch the gorgonian at a fork in a branch, all the polyps on both sides of the fork, and beneath the fork, will retract. (This is clearly not plant-like behavior!)

The first time I saw true soft corals was on Australia's Great Barrier Reef, and I was fascinated by the lovely formations and particularly by the saturation of colors: purples, reds, oranges, pinks, white, blues. Rainbowed cauliflower corals, I thought. And always, upon close examination, the soft

The lavender color of this branched coral is not seen in the Caribbean but is common in the Coral Sea.

Fire coral in closeup.

corals show the polyps characteristic of all corals.

For divers, no discussion of corals would be complete without a section on fire coral (although, technically, fire coral isn't a true coral). It is certainly a coelenterate, though, and it certainly has nematocysts that get the attention of a victimized diver!

Fire coral is mustardy yellow in coloration; some types are pale mustardy yellow and some a more rust-colored mustardy yellow. The stings of the darker-hued fire coral hurt more than the stings of the pale stuff — I speak from experience. The best way to learn to identify fire coral is to have someone point it out for you; draping your bare skin over various mustardy colored formations would give you the same information in a much less comfortable way. Careful observation can also help: the surface of fire coral looks smoother than the convoluted surfaces of the true stony corals.

Another difference between fire corals and stony corals is the way they grow. Some fire corals grow in vertical plate-like formations in very shallow waters, but most fire coral is found encrusting other reef structures: dead sections of gorgonians (such as those eaten away by flamingo tongues), dead stony coral heads, abandoned anchor lines, etc. In contrast, each species of true stony coral has its own characteristic formation.

One interesting thing about fire corals that many people never discover is that despite the real pain that their sting can cause to vulnerable skin, nematocysts can't penetrate the palm surfaces of a person's hands. The skin there is thicker, and protects us from fire coral and many other nematocyst-bearing critters, though I've never been brave enough to test my hands against, for instance, a man 'o war's tentacles.*

*PADI doesn't recommend you try it either!

Jellyfish

A jellyfish is an upside-down (or sideways) anemone-type polyp. Even more than anemones, jellyfish are known for their stinging potential. Since jellyfish are pretty much free-floating animals, living in the open oceans of the world and usually arriving on coastlines by accident, they're not of much interest to divers except as critters to avoid.

One exception to the nomadic habits of jellyfish is *Cassiopeia xamachana,* the upside-down jellyfish, which lives in great numbers in the waters of the Gulf side of the Florida Keys. This jellyfish sometimes swims in normal jellyfish position, that is, mouth down, but it is more commonly found settled on the bottom, mouth **up**. *Cassiopeia* is also unusual in that, like the stony corals, it often has commensal zooxanthellae living within its body; thanks to the nutrients that the zooxanthellae provide through photosynthesis, the upside-down jellyfish can rest on the bottom for long periods of time without having to actively search for food. People often wade among upside-down jellyfish without ever realizing they are jellyfish, but sometimes *Cassiopeia* reveals itself by inflicting painful stings.

The most highly venomous jellyfish is the sea wasp of the Great Barrier Reef, which has killed human beings in a matter of minutes. On my first Australia trip I was puzzled by the fact that the crew of the boat seemed totally unconcerned about sea wasps, but later I learned that sea wasps are most prevalent in inshore areas during Australian summer months. In two trips there (during September and October) I never saw one.

The Caribbean has its own sea wasp, but (like most Caribbean relatives

It's quite unusual to see Caribbean sea wasps in the daytime, but when we found this one I couldn't resist touching one of its tentacles with a piece of hot dog. The sea wasp stuck to the hot dog and drew it into its body cavity! Photo by Ed Cieiakie.

of Indo-Pacific venomous creatures) it is not nearly so dangerous as the Australian sea wasp. It's almost always seen at night in areas lit by bright lights. Sea wasps generally stay at the surface and are thus easily avoided: divers are advised to descend as soon as they enter the water and ascend and exit directly at the end of their dives if sea wasps may be in the area.

I did have an interaction of sorts with a Caribbean sea wasp one strange afternoon when we'd had a wind shift and the water was quite cloudy. Our group of divers was heading for the drop-off from shore, and in about ten feet of water over a sand bottom we found a sea wasp. We all looked at it — a clear, cup-shaped bell with flattened sides and four trailing tentacles — then looked over our shoulders to make sure it was alone, which, as far as we could see, it was. Sea wasps can swim, but not so well that a diver can't easily stay away from them (as long as the diver knows they're there).

Emboldened by my superior swimming ability and in possession of my usual fish food, I took a piece of hot dog and held it very cautiously against one of the sea wasp's tentacles. The tentacle stuck to the hot dog, and as we all watched, the sea wasp drew the hot dog slowly to its mouth and ate it! While this was going on I put a piece of fish on another tentacle, but the sea wasp dropped it! Did it drop the fish because it preferred hot dog, or because its attention was already fully taken by the hot dog, or. . .?

Eventually we continued to the drop-off, leaving the sea wasp with the hot dog visible through its clear body as evidence of the meeting of ancient and modern life.

Another jellyfish-type animal that tropical divers are likely to see occasionally is the Portuguese man-of-war, which is actually a colony of polyps rather than a single polyp like most jellyfish. The nematocysts on the man-of-war's tentacles are extremely potent, and can fire regardless of whether they're attached to the bubble, stranded on shore, or extended thirty feet or more away from the bubble. I once saw a screaming swimmer rescued by lifeguards and rushed to the hospital as a result of a tangle with man-of-war tentacles.

One handy thing is that the bubble float, which keeps the man-of-war colony at the surface of the water where it is pushed around by the wind, cannot sting — so if you must handle a man-of-war, do it by the bubble. (Grabbing a man-of-war carefully by the bubble is a reasonably safe way to impress most of the people on the beach.)

Strangely enough, there is a small banded fish called the man-of-war fish that lives among the tentacles of the man-of-war and is thought to eat scraps left over from the man-of-war's meals of fish — yet the man-of-war fish isn't immune to its protector.

The man-of-war isn't the only coelenterate that gives shelter to other animals. Various fish, shrimp, and crabs all live their lives within or under the protection of the tentacles of anemones. A diver could do worse than spend a series of dives studying the coelenterates and their roommates.

WORMS
(Phylum Annelida)

Wait! Okay, some worms are yucky, but don't skip this chapter! Worms can also be interesting, even beautiful. Besides that, this is a short chapter and it deals mostly with the worms divers are most likely to see: bristleworms, fan worms, and what just may be your own night-diving mystery creature.

Unlike the sponges and the stingers, worms live virtually everywhere: on land, in fresh water, as parasites in the bodies of other animals, and, of course, in marine environments.

Annelid means *ringed,* and a major characteristic of the phylum is repetitious body segments. Each segment has separate but essentially similar systems. Also, each body segment has a pair of *parapods* (side feet), which can be used for locomotion, and which, in the free-moving marine worms, contain the gills.

Most marine worms are in the Class *Polychaeta,* which means *many bristles.* The first body segment of polychaetes is different from the rest of the segments in that it contains the worm's sensory equipment: light-sensitive eyes, antennae, and receptors called palps for giving chemical and tactile information. The second segment contains the animal's mouth.

The polychaetes are divided into two major subclasses: the sedentary worms (Subclass *Sedentaria*), and the free-moving worms (Subclass *Errantia*). (This distinction stimulated me to look up the word *errant,* as in knight-errant. The first definition is "misbehaving," which may have fit the knights but seemed wrong for the worms. The second definition was more satisfactory: "traveling in search of adventure.")

Errant Polychaetes

The errant polychaete that divers know best is the fire worm, also called the bristleworm. Bristleworms have the characteristic segmented bodies

A feeding bristleworm. Its bristles are probably extended because it was disturbed by the framer.

of the annelids, yet they look caterpillar-like because the bristles act like little legs. Bristleworms are beautiful animals, with white bristles (*setae*) on their parapods and orange or green bodies. Like most errant polychaetes, bristleworms are carnivorous. One of their favorite foods is coral (including fire coral). They are often found engulfing the tips of staghorn coral branches with as much as two or three inches of the branch within their bodies. In this position, the bristleworm secretes its digestive juices on the coral polyps, then absorbs the partially digested meal. All that's left is the completely white tips of the corals — the color is gone because the polyps have been eaten.

Bristleworms also eat anemones — some have eaten anemones ten times their size!

Bristleworms are scavengers, too. Any dead fish that's been lying on the bottom will soon have several bristleworms firmly affixed to it. When a disease hit Bonaire's long-spined urchins in 1983, many half-dead urchins stumbling around in the late stages of the disease were already being fed upon by bristleworms.

In the course of my "Dee feeds the world" diving adventures, I've experimented successfully with feeding bristleworms. A hungry worm raises its head when it smells the fish; the worm moves rapidly to the food, opens its mouth, and engulfs the meal. It's a fascinating process to watch.

Something always struck me as weird about this bristleworm method of feeding, and I finally figured out what it was: Whereas most animals —

humans included — move the **food** into their mouths to eat, the bristleworm moves its **body** onto the food.

Once a crawling bristleworm moved onto my fish-scented hand and
tasted me several times — it felt like being licked by a cat — before realizing that the fish smell didn't extend past the surface of my hand, and it
wasn't going to get a meal. It moved away in search of better hunting. Isn't
it a relief to know that these carnivores aren't interested in human flesh?

Bristleworms are best known to divers for their defense mechanism. Touch a bristleworm bare-skinned, and you instantly understand
the name "fire worm." Those beautiful white setae on each parapod are
poisonous, and being zapped results in a throbbing, burning sensation
that lasts an hour or so. The bristles can sometimes be removed by sticking tape over them and pulling it off, but the pain (since it's caused by venom that doesn't come out with the setae) persists for a while anyway. The
bristles can also be imbedded in gloves or a wet suit and then transferred
to skin, and they hurt almost as much when encountered secondhand.

Like most marine animals, bristleworms would much rather go
about their business than hurt divers; they only retaliate if they're dis-
tressed. When the bristleworm crawled up onto my hand, it didn't hurt
me; to the worm, my hand was just a surface with an interesting scent. On
another occasion, though, a bristleworm investigating my hand walked
into my wet suit cuff, frightened itself, and bristled onto my exposed wrist.
That hurt!

I saw a bristleworm use its defenses quite effectively on a little
grouper once. I'd been feeding the worm a piece of fish, and a greedy
coney zoomed up and gobbled down the fish — and the bristleworm. The

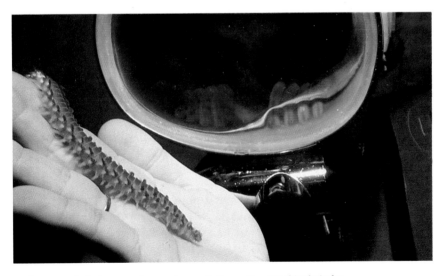

As long as a bristleworm is not stressed, it won't extend its bristles.
Photo by Rod Canham.

coney instantly spit out its booty and swam off, holding its mouth open and shaking its head from side to side.

Reproduction in errant polychaetes is associated with the cycles of the moon. In some species, at spawning time the worms' parapods develop for swimming (rather than walking) and the eyes become enlarged, and all the worms swim to the surface of the water. The males release sperm, which causes the egg-bearing segments of the females to burst. Fertilization occurs in the water. The eggs enter the plankton layer and the adult worms return to the bottom.

One particular polychaete called the Bermuda fire worm spawns each summer for the few days following the full moon. At spawning time the normally bottom-dwelling worms rise to the surface where each female swims in a tight circle and gives off a bioluminescent substance that attracts males. The green circles of light can be seen from the surface; this worm's spawning activities may have been what Columbus reported as "candles moving in the sea."

I saw something that may have been Bermuda fire worms on a dive trip in Belize one August near the time of a full moon. I didn't notice any "undersea candles" from the surface, but on night dives I did see bioluminescent wormy things moving in frenzied corkscrews. When I flicked my fingers at them they seemed to explode — and what was left looked like a small bristleworm in a milky cloud!

The best-known polychaete spawning activity is that of the Samoan palolo worm, which spawns at dawn one week after the October and November full moons. The sex cells develop in the posterior portion of these worms, and at spawning time the posteriors break off and rise to the surface. There are so many of them that the natives (who consider them a delicacy), scoop them up in buckets. When the sperm and eggs are released from the worm segments the entire surface of the sea turns milky. The worms regenerate their posterior segments in time for the next year's spawning.

Sedentary Polychaetes

Sedentary polychaetes can't search for adventure — it must find them. They are unable to move around because they live permanently in tubes.

Imagine a bristleworm in a tube with only its head sticking out, its tentacles transformed into branched "flowers," and its unused parapods, within the tube, degenerated. Your mental picture is that of the sedentary polychaetes most familiar to divers and snorkelers, the filter-feeding worms.

These worms don't need and don't have jaws, since they eat small particles they filter out of the water. The filtering mechanism, called the *radiole,* is tentacle-like in structure, and each branch (*pinnule*) is lined with tiny hairs (*cilia*) that create a water current. Food particles carried by the water are funneled into grooves along the pinnules and into the mouth of the worm. The worm also gets oxygen from the water it filters.

The side of a Christmas tree worm is an interesting perspective.

Eyespots on their radioles warn tubeworms of shadow, which generally causes them to withdraw their radioles. A strong current, or jarring the base of the tube (such as with a camera's extension tube framer) causes the same reaction.

What is most wonderful about these worms is not how they feed or breathe or move, but how they look. The radioles are delicately structured and beautifully colored.

One of the most commonly seen tube worms is the Christmas tree worm, impossible to miss on any tropical reef: two matching red or orange or yellow or white whorls, looking like little Christmas trees, at the entrance of each worm's tube. It's almost impossible not to reach toward them — which usually results in the whorls disappearing as the worm retracts.

Close examination shows that the end of the tube is blocked off by the worm's *operculum* (trap door) and protected by a sharp spine. Nothing of this worm is ever seen by divers except its radioles, its operculum, and the outside of its tube. Its body is trapped in the tube for life.

Sedentary worms are divided into two families: the *serpulids* (such as the Christmas tree worms), whose tubes are calcareous, and the *sabellids*, whose tubes are non-calcareous.

Many sabellids are even more light- and movement-sensitive than the Christmas tree worms — sometimes it seems as if a diver has only to look at one of these worms to cause it to retract into its tube. The tubes of sabellids are more apparent than those of serpulids; often an inch or more of sand-grainy tube protrudes after the worm retracts. Sabellids have no operculum; they retreat far enough back into their tubes for safety from predators.

This is the only time I've ever seen more than one color of worm in such a cluster.

One of the most beautiful sabellid worms is a type that grows in clusters — bouquets — on some Caribbean reefs. I have seen them colored either orange or lavender, and on a very few occasions have seen worms of both colors in the same bouquet. Their beauty is enhanced by their behavior: they are unusually reluctant to retract into their tubes, so it's easy to photograph them. They're common in the Bahamas, but I didn't see any on Bonaire's reefs until I dived the east (windward) side of Bonaire. Perhaps Bonaire's calm waters don't present enough challenge for these worms!

Although filter-feeding tube worms are sensitive to light and shadow, they feed at all times of day. The lack of movement of a sleeping fish is sometimes evidenced by expanded tubeworm radioles in contact with its body.

A third family of sedentary worms, the *terebellids,* are a night diving mystery: "What **was** that spaghetti-like stuff we saw along the sand that was sucked into a hole when we touched it?" The strands are the tentacles of a terebellid worm, extended along the bottom in search of the organic deposits — detritus — on which the worm feeds. A terebellid can move particles to its mouth by the action of cilia on its tentacles or by retracting the entire tentacle with the food particle. It breathes through gills located near its mouth. Terebellids can regenerate lost portions of their tentacles or gills, a skill I suspect they have to use frequently.

Like their errant cousins, sedentary worms have separate sexes. They release eggs and sperm into the water, where fertilization takes place and planktonic larvae develop.

Many marine worms live in burrows under the sand, or within sponges or corals, and when unearthed appear just as unappealing as their terrestrial relatives. The marine worms most likely to be seen by divers, on the other hand, are among the most lovely creatures on the reef.

MOLLUSKS
(Phylum Mollusca)
outrageous diversity within shells

It boggles my mind that animals as widely diverse as a clam, a snail, and an octopus are related, yet they're all members of the Phylum *Mollusca*. The statistics on this phylum are interesting: there are over 125,000 species of mollusks, ranging in size from less than an inch to over 50 feet in length; some live on land, some in fresh water, and most of them in the sea — from the surface to depths of more than two **miles**!

This unwieldy phylum is divided into three major groups of interest to divers: 1) *bivalve* (two-shelled) mollusks, such as clams and oysters; 2) *univalve* (single-shelled) mollusks, also called *gastropods*, which we commonly think of as snails (although this group includes the shell-less mollusks [nudibranchs] and those with vestigial shells like the sea hares), and 3) the *cephalopod* (or "head-footed") mollusks, the octopuses and squid.

Since these varied critters are all in the same phylum, they have a lot in common. In general, all mollusks have true organs, including a heart, stomach, intestines, kidneys, and gonads; they have a mantle (fleshy tissue that surrounds their organs); they have a muscular foot, usually used for propulsion; and they are unique as a group in possessing the radula, a tongue-like structure covered with fine teeth that can be used for scraping, grasping, drilling, or cutting, depending on its owner.

Many mollusks are missing one or more of the characteristics they are all supposed to have — perhaps the only truly common characteristic of the mollusks is that they defy generalization . . . In this spirit of confusion, then, please overlook my use of qualifiers such as "probably" and "most."

Bivalves

The most familiar of the mollusks are the bivalves: clams, oysters, and scallops, for instance. Bivalves are named for their two shells, which are connected by a hinge. The bivalves demonstrate the unwillingness of the

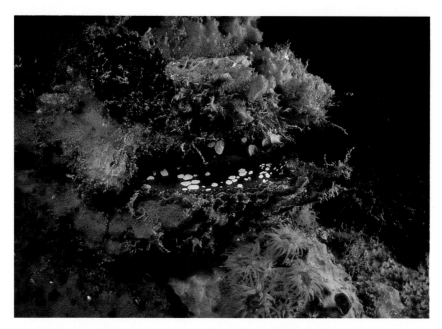

The larger and older the scallop, the more encrusted its shell is likely to be.

mollusks to conform to any rules by **not** possessing that "characteristic of mollusks," the radula. They don't need radulas, though, because bivalves feed by filtering small particles from the water. They obtain oxygen, too, from the water they filter. One experiment found that an average-sized clam filters almost three quarts of water an **hour.**

One reason that giant clams are so readily poached is that they often grow like this one: resting on the bottom on its hinge, not attached to anything.

Most of the 20,000 or so species of bivalves live on or beneath the substrate and are not commonly seen by divers. This lifestyle, plus their double shell, provides them with protection from predators. A notable exception to the bivalve's general unobtrusiveness is the giant clam of the the Pacific, which, growing to as much as four feet across, is the largest of the bivalves. Giant clams have the reputation — in fiction, at least — of spending their free time plotting ways to grab skin divers and hold them under water until they drown. The truth is that since they're filter feeders, giant clams don't have any reason to capture people, and in fact many of them are so encrusted with other life that they can't even fully close their shells.

Thus a diver in giant clam waters is, rather than endangered by these animals, privileged by the sight of their beautifully colored mantles decorating the bottom of the sea. Living in the mantles of giant clams are commensal zooxanthellae, whose pigment gives the mantles their lovely coloration and protects the clams from sunburn.

A Caribbean bivalve familiar to observant divers is the flame scallop, alias the lima clam, rough lima, and even *Lima scabra*. This critter is generally found in reef crevices, sometimes attached by a secretion called *byssal threads*. The flame scallop's mantle has beautiful red or white tentacle-like extensions, and often its body is pure red in color. Because it prefers reef crevices, the flame scallop is a photographer's frustration — the ledge usually prevents insertion of an extension tube.

Once in a while, though, a flame scallop is found in the open and photography by a patient practitioner is possible. At first the scallop closes completely. This is the time to get the camera system in position. Ever so slowly, the shell begins to open, and the "tentacles" unfold one by one.

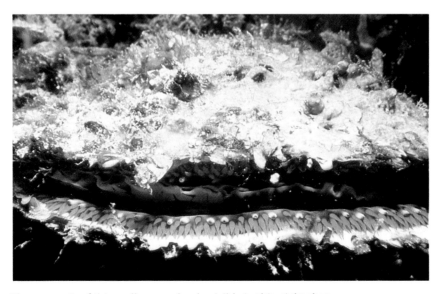

The eye spots of this scallop are clearly visible in this night shot.

Finally the shell opens fully. Usually then — just before the picture is taken — the scallop forcibly snaps its shells together, zips backwards a few feet and clams up again.

This lesson teaches the photographer the proper moment to trigger the shutter next time; the benefit of the lesson is a great photo of an unusually designed and colorful marine bivalve.

Some bivalve mollusks reproduce much like sponges; that is, males release sperm into the water where it enters the incurrent openings of females; eventually larvae are released into the plankton layer. With other bivalves, such as oysters, both eggs and sperm are released into the water, and the eggs are fertilized by chance and proximity. Immature bivalves go through various larval stages before they settle down (literally) to their adult shapes. Their success as adults is dependent on the final stage planktonic larvae finding a hospitable environment in which to live.

Gastropods

There are over 100,000 species of gastropod mollusks, which makes them the largest subdivision of the phylum. In general, they have a single shell, and most of us grew up calling them *snails.* Despite the protection that their shells provide, gastropods are not satisfied with the shell as their only protection, and are often either surprisingly well-camouflaged, well-hidden, or both. Whenever I'm in danger of becoming overly proud of my ability to find animals under water, I need only to flip through a Caribbean shell guide to be humbled: in more than 5000 dives in that sea, I've never seen hundreds of different species of resident gastropods.

Most single-shelled mollusks have a foot upon which they walk or glide across the bottom, simple eyes that can detect light intensity, and a coiled shell that protects their internal organs. Unlike the filter-feeding bivalves, the gastropods are algae-grazers or carnivores, and possess a radula.

Aside from the few gastropods (like the queen conch) used as food, people's interest in these mollusks usually extends only to their shells — cowrie shells have been used as money by some societies, and everyone appreciates the beauty of seashells. What most shell collectors don't even realize they're missing is the fascinating animals who create those shells.

Take the nerites, which live in the littoral (tidal) zone. When the rocks they live on are covered with water, nerites feed on the algae covering the rocks. When the tide recedes, each animal retreats into its shell. It closes its *operculum* (a horny substance that fits perfectly into the shell's opening like a trap door), seals the opening to retain moisture, and waits for the tide to come in again. Nerites can wait for water for incredibly long periods of time. I once tried to keep some in an aquarium, and they persisted in crawling out. I'd have to find them and put them back into the tank almost daily, so I finally released them where I had collected them. About **two months** later, I found three nerites I had missed hidden in a

A conch shell, with one stalked eye protruding forward.

corner. When I placed them on a wet rock the animals immediately
emerged and began to graze! Littoral zone residents have earned their
reputation for hardiness.

The queen conch feeds on sea grasses and other marine algae,
usually in shallow waters. Since it generally moves slowly on top of the
sand, various marine plants and even animals — occasionally some
encrusting sponges or corals — grow on its shell. A snorkeler hunting for
queen conch from the surface may have trouble differentiating the
conch's shell from all the other rocks littering the bottom — until the shell
is turned over and the lovely sunset-colored lip is revealed. Regardless of
the amount of growth on the top surface of the shell, the lip is always
clean; the conch's mantle, which secretes the shell, protects the lip.

At one end of the conch shell is an upturned area through which
the animal's eye protrudes as it feeds. At the end of the conch's body is the
claw-like operculum. Unlike the nerite, the queen conch can't completely
cover its shell's opening with its operculum, because the operculum is
shaped for another purpose: escape. When the conch needs to move
quickly, it hooks its operculum into the bottom and vaults away from
predators such as starfish.

Another gastropod with a claw-like operculum is the tulip shell, a
fairly common animal in the shallow waters on the Gulf side of the Florida
Keys. The tulip shell is smooth in texture and either light or dark brown
with contrasting fine lines. The body of the animal is, surprisingly, red.
Another unexpected characteristic of the tulip shell is that it is a carnivore,

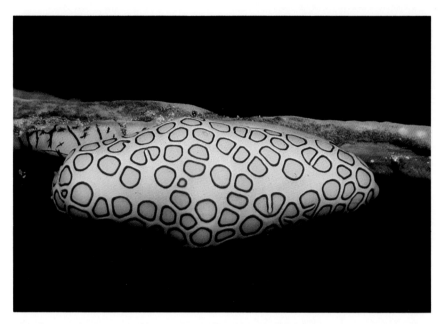

This flamingo tongue snail has eaten the flesh from the gorgonian it is crawling upon.

and a cannibalistic carnivore at that. Snorkeling off the Florida Keys, I saw one tulip chase another, both of them vaulting along on their claws. The next day I found the clean, empty shell of the smaller of those tulips.

A friend of mine discovered the tulips' carnivorous nature much to her chagrin, after she'd placed a live tulip shell into an aquarium that already had a population of flamingo tongue snails. The next morning all that was left of the flamingo tongues was their shells, and she perceived the tulip looking very smug.

Her loss was particularly distressing since the live flamingo tongue is much more beautiful than its empty shell. When alive, the animal's mantle — light in color with orange-yellow spots outlined in black — covers the shell in a leopard pattern. With the animal gone, the shell is a less-interesting pale orange in color.

Living flamingo tongue shells are almost always perfect and are never encrusted with any growth. This perfection is because the mantle covers the shell most of the time, protecting it and also constantly secreting more material to thicken it. The markings on the flamingo tongue's mantle are thought to provide camouflage as the animal grazes on the gorgonians that are its major food source — but, at least for divers, the brightly-spotted mantle makes the shell easier to find.

It takes a very sharp-eyed diver to spot a cowrie, though, no matter how out in the open it may be. That's because the cowries, close relatives of the flamingo tongues, camouflage their shiny shells with their dark, bumpy mantles. Cowries are protected from their marine animal predators by their thick shells, but it's their mantles that hide them from human predators!

Other gastropods have unencrusted shells because they spend part or all of their lives beneath the sand. Olive shells move around under the sand in the daytime, leaving characteristic trails. When I'm lucky enough to be able to find the end of a trail (which is often more difficult than you'd think), I dig out the shell. Naturally the animal retracts into the shell as it's abruptly pulled from under the sand, but eventually it emerges and digs down into the sand again. The front part of the olive shell's body is arrow-shaped, a good digging shape, and the animal can completely rebury itself in less than 35 seconds.

In Australia I had great fun finding lovely auger shells as long as six inches beneath the sand — until I unexpectedly pulled up a cone shell instead. Most cones are venomous and some of the Australian ones are as dangerous as any in the world, so I had to quit that game. . .

Many gastropods will leave their daytime hiding places at night to forage openly. Olive shells often crawl over the surface of the sand at night. Flame helmet shells, a small (up to about 5 inches) variety of helmet shell with a vibrantly orange and black lip, also emerge to the sand's surface at night. I've never seen a flame helmet in the daytime, but because only the very tops of their shells ever have any growth on them, I believe that they spend their daylight hours almost completely under the sand. A sharp-eyed diver can find all kinds of small gastropods on the reef at night, usually on the algae-covered areas rather than on the living corals.

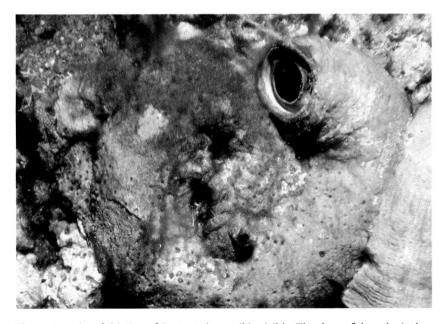

The entire tube of this Sea of Cortez tube snail is visible. The door of the tube is the animal's operculum.

Another type of mollusk that is likely to be around algae-covered areas of the reef is the worm snail, which I call tube mollusks. Only the tubelike end of their shells is visible, but looking into it reveals the face of a critter very much like any other gastropod mollusk. This animal can't move its shell, however, because it's fixed firmly in place. The gastropod feeds by spinning mucus strands into the surrounding water to trap tiny organisms, which it then delivers to its mouth by retracting the mucus. Most tube mollusks have no operculum, but they can pull their bodies far enough into their tubes to be safe from predators.

The sexes are separate in gastropod mollusks, and after internal fertilization takes place the female lays eggs. Generally the eggs of gastropods are protected in some way, from the sand collar (eggs and sand grains held together by mucus) of the moon snail to the egg capsules of whelks and some murex shells. Some of the young hatch as larvae and enter the plankton layer to develop further and finally settle as juveniles; other young skip the larval stage and hatch as fully-formed juveniles.

MORE MOLLUSKS
outrageous diversity without shells

The wealth of information available about shelled mollusks is primarily due to the shells themselves, because they survive the animals and because of their interest to collectors. Shell-less mollusks have attracted much less notice from the world in general.

Nudibranchs

For our purposes, any shell-less mollusk that is not a cephalopod is a nudi-branch, although technically there are several defined groups of these animals.

Because embryonic nudibranchs have coiled shells, they are classi-fied with the gastropod mollusks; but they're different enough from other gastropods to deserve separate discussion here.

Those who are familiar with nudibranchs know that they surpass in beauty even the most lovely of shelled gastropods. I've had exactly 13 dives off the California coast, and my outstanding memory of those dives is the diversity of shape and color of the nudibranchs there. There are pink and scarlet nudis, white and orange nudis, blue and red nudis. An interior decorator would have no trouble finding a California nudibranch to match any color scheme!

Like other gastropods, nudibranchs have a pair of tentacle-like pro-trusions on the front of their bodies called *rhinophores,* which are used for detecting odors, and which can be retracted into their bodies for pro-tection. The name *nudibranch* means *naked gills*, and although nudibranchs breathe through their body surfaces, they usually have some kind of gill structure as well.

Some nudibranchs are herbivores, using their radulas to scrape algae from the surfaces of rocks; others are carnivores, eating hydroids,

This two-inch or so Coral Sea nudi has a particularly high gill cluster.

corals, and anemones.

One group of nudibranchs, the dorids, feeds on sponges, tunicates, and bryozoans. Each member of this group is characterized by a whorled gill cluster surrounding its anus. A dorid nudi can retract both its rhinophores and its gill cluster.

Probably the best known dorid nudibranch is the Spanish dancer of the Red Sea. Unlike most nudibranchs, which grow no larger than a couple of inches, the Spanish dancer can grow to twelve inches or more in length. It is so large that it often has a pair of shrimp living around its gill cluster. When this nudi is released into mid-water it undulates its body like its namesake, and "dances" its way to the bottom. If it lands in an inhospitable place it will dance away, and continue the process until it lands somewhere acceptable.

The Spanish dancer "dancing."

A close-up of the rhinophores of the Coral Sea sea hare.

Even more fantastic in shape than the dorid nudis are the aeolid nudibranchs, which have, instead of a single gill structure, a series of growths all along their dorsal surfaces, called *cerata*. When an aeolid nudi eats coelenterate polyps, the nematocysts of their prey somehow pass **un**discharged into the aeolid's cerata, where they **will** discharge against predators of the nudibranch!

Sea hares are very similar to nudibranchs, but they generally grow larger than nudis do, and sometimes they have a thin internal shell. They're named for rabbit-ear-like sensory projections, and perhaps also because they are voracious herbivores. Two wing-like flaps — extensions of the animal's mantle — curl upward along the sea hare's sides, and some species have a gill concealed at the top of one of them. Sea hares, like their cephalopod relatives, eject a purplish-red ink when they are disturbed.

Like all mollusks, nudibranchs and sea hares have multiple defense mechanisms: the aeolids have transferred nematocysts; nudis with bright colorations "disappear" among lots of other small brightly-colored animals (such as tube worms and encrusting sponges); some nudis and many sea hares perfectly match the animals and plants on which they feed; many nudis (including the Spanish dancer) spend the daylight hours hidden in crevices in the reefs. In addition to all this, nudibranchs and sea hares are distasteful to fish.

Just how distasteful nudis can be was demonstrated by Bonaire's most common nudibranch, the ruffleback nudi (*Tridachia crispata*). Usually light blue and white in color, this critter reaches a maximum size of about two inches, and feeds mostly on algae. It makes no effort to hide, but

A closeup of the feathery gill cluster of the leathery nudibranch.

is small and not particularly noticeable anyway. I'd never seen any predator pay it the least bit of attention, until the time that a ruffleback nudi was
tossed by accident into the water column near a mob of Bonairean
"attack" yellowtails. The yellowtails zoomed toward what they assumed
was a tasty morsel, but soon braked frantically and swerved away. A young
one finally sucked down the nudi — and spit it out instantly — unharmed!

In addition to their efficient defense mechanisms, nudis have efficient reproductive methods. Most individuals have both male and female
reproductive organs. They don't fertilize their own eggs, though; when two
of these usually solitary critters meet, each fertilizes the eggs of the other,
thus doubling the reproductive efficiency of each mating meeting. The
eggs are probably just as distasteful as the nudis, because although they're
unprotected by armor or camouflage, they're ignored by predators. When
the larvae hatch, they enter the plankton layer like those of many other
gastropods.

Cephalopods

Flame scallops are wonderful to photograph, gastropod mollusks are a
challenge to find, and nudibranchs are a treat to the eye, but the cephalopods — squid, cuttlefish, and octopus — to me are the most fascinating
members of the Phylum Mollusca because of their superior intelligence.
Their behavior is more interesting than that of any other invertebrates —
perhaps because it is determined by reasoning rather than instinct.

There are over 100,000 species of gastropods and 20,000 species of
bivalves, but fewer than one thousand species of cephalopods, all of

A close-up of the egg mass of the California sea hare.

which are marine animals.

Like (most) other mollusks, cephalopods have a radula; a mantle, which protects their internal organs; and a foot that has developed into the arms and tentacles that surround their mouths. Unlike other mollusks they have a highly developed eye. As a matter of fact, the eye in cephalopods is very like the human eye.

This octopus seems to be squinting at the light from my strobe. Little Cayman, 1:3.

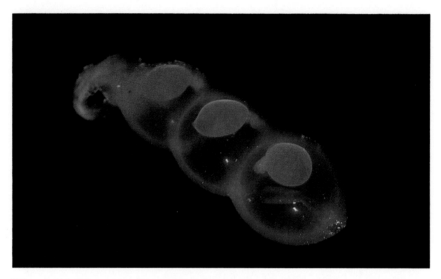

When I took this macro photograph I didn't know what my subject was; a close look now reveals a tiny cephalopod in the foreground chamber.

Cephalopods are also the only mollusks who can change color, and the only animals who can control their color changes by muscular action. They have numerous pigment sacs of different colors, called *chromatophores,* which they can expand or contract almost instantly. As a pigment sac expands, its color appears; as the sac contracts, the color disappears. Cephalopods can change color at will, or even cause waves of colors to pass over their bodies.

Squid

The squid are the best swimmers of the cephalopods; whales have been found with squid sucker marks on their bodies, which testify to both the size of some deep-ocean squid and the squid's swimming ability.

Squids move by jet propulsion, sucking in water around their mantles and ejecting it forcefully through their funnels. They can move with equal ease forwards or backwards by changing the direction of the funnel, and fine-tune their movement with a single fin on each side of the body. Whether they are moving or not, though, squid breathe by circulating water constantly through their mantles and out the funnels.

Reef squid spend much of their time hovering in mid-water or close to the surface, where they can be seen by snorkelers as well as scuba divers. The first time I saw a squid it registered in my mind as a sort of broken fish, since its outline had a little bend where the tentacles met the body. Often squid will hover in formation like a group of hummingbirds, gently moving their stabilizing fins, unafraid and clearly interested in the activities of marine *Homo sapiens.* I've never been able to get closer to them

Occasionally on Little Cayman night dives, we saw large aggregations of tiny squid. Photo by Lloyd Garland.

than a few feet, and I always feel that they maintain their distance because I'm not interesting enough.

Squid have eight arms and a pair of longer, extendable tentacles, which they use for capturing prey. One night the lights of a research vessel in the shallow waters of the Bahamas Bank attracted a big school of minnows, which in turn attracted a group of squid. Several of us entered the water with masks and snorkels and were able to watch the squid feed: they hovered near the school, and every so often one would dart forward and grab a fish with no more effort than we would use to select a piece of fruit from a display!

One fascinating day I was privileged to see squid mating, a day in which I did three dives in the same location. On the first dive I saw two squid that looked as if they were paying an unusual kind of attention to each other, but an eager photographer disturbed them and they left the area. On the second dive I managed to signal to my buddies that the squid might be mating and we watched for a bit, but it was the third dive that really rewarded us, since by then I was pretty sure what was going on and was able to prepare my buddies before we entered the water.

We found the squid almost immediately. They were quite close to each other, but not parallel as squid usually are when they're hovering. They changed colors rapidly, and, as we watched, they swam together with the male just below the female. He extended a specialized arm — called his *hectocotylized arm,* to which he had already delivered spermatophores — above him and into the mantle cavity of the female. The two squid remained like this for just a few seconds and then separated.

The female continued to change colors, then swam along the bottom investigating likely ledges under which to deposit her eggs. Finally she hovered for a minute or so with all her arms pointing beneath a ledge.

During this time, the male watched her impatiently. ("Dear, you're spending an awful lot of time with the children; please don't forget **me**!")

When we peeked under the ledge we could see the egg capsules of the squid.

This pair continued their matings for quite a while, and the female deposited each cluster of eggs in a different place. Since the egg clusters are abandoned after they're deposited, the many different sites increases the likelihood that at least one cluster will survive predation.

Other species of squid, such as the common Pacific squid, mate in huge congregations. The Cousteaus documented such a congregation, and at the same time got pictures of sharks gobbling up the preoccupied squids.

Mating squids are probably the only squids that predators have a chance to get, because their speed and intelligence make them difficult prey. A squid has a trick up its mantle, too: a squirt of ink that keeps its squid shape in the water and distracts predators from the real thing.

Octopus

Octopuses ink, too, but their method is different than the squids'. First the octopus turns dark, then it expels its dark-colored ink and almost simultaneously turns a lighter color. The octopus' ink doesn't necessarily look like an octopus, but it holds the predator's attention so the octopus is able to escape — or to "disappear" by mimicking the color and texture of whatever it rests on.

Inking is not the octopus' only means of defense, of course. Although octopuses usually walk along the bottom on their arms, they can also swim quickly (though not as quickly as squid) through the water using jet propulsion. They can ooze into crevices in the coral, and I've seen — luckily more than once, because I didn't believe my eyes the first time — octopuses actually bury themselves in sand!

One particular octopus, the blue-green octopus commonly seen at night on Bonaire, displays a fascinating escape behavior that I've never seen described — at least, not in relation to octopuses. When disturbed, this little octopus gracefully "climbs" an invisible ladder into the water column using two or three arms, with the rest trailing behind — instead of the usual all-arms-behind-and-full-speed-ahead escape method — until it's 15 feet or so above the reef. Then it jet-propels itself back to the bottom away from whatever has frightened it.

This method of escape parallels that of the flying fish, which leaps out of its environment (the water) and also randomly returns, hoping to elude its pursuers in the process. The blue-green octopus also leaves its environment (in this case, the reef bottom), and also randomly returns to it. If the octopus was trying to elude a hunting moray eel, for example, this escape method would be quite effective — especially for an octopus in the dark, away from its den.

An octopus **in** its den has even more methods of defense. One time the fish I had fed an octopus attracted a moray eel. As the eel investigated the octopus's doorway, the octopus aimed its siphon at the eel and blew out a strong jet of water. The eel retreated in surprise, but only momentarily. The octopus changed its defensive tactics by calmly rolling its tentacles back and presenting its under-surface — all suckers with mouth and beak in the middle — to the eel, who was unable or unwilling to bite through the tough surfaces of those suckers. As the eel began to swim away, the octopus reached out and grabbed it! The eel dashed off and the octopus calmly resumed its snack.

Octopuses are said to eat shelled gastropods and crustaceans, and crab and mollusk shells often litter the front yards of octopus dens — at least the dens of the common octopus of the Caribbean. Each discarded gastropod shell has a tiny hole drilled near the top of the whorl; the octopus drills the hole with its radula, envenomates its prey, and then pulls the animal out of the shell and eats it. The common octopus usually hunts in low-light situations such as dawn and dusk, but is rarely seen out of its den.

I've never seen the blue-green octopus in a den — as a matter of fact, I've never seen this octopus under any other circumstances but out hunting at night.

The blue-green octopus is smaller than the common octopus, with its body-head about the size of a cup or a little bigger, and it has long arms with a delicately thin webbing between them. To hunt, it travels along the bottom with its arms extended around it, occasionally dropping down

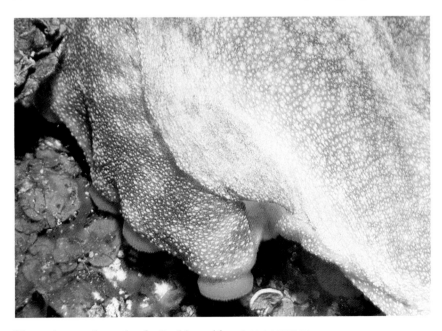

The suckers and mantle of a Caribbean blue-green octopus.

over a rock like an open parachute. Then, it curls some arms under the webbing and feels around for trapped animals.

One night I was watching the blue-green octopus hunt, when suddenly its webbing started twitching as if the octopus were having some sort of seizure! When the twitching stopped I was able to look beneath the little octopus — and in its mouth was a newly-captured damselfish!

Usually when I watch a blue-green octopus hunt I try to shine only the dim edge of my light beam on it, to disturb it as little as possible, but the next time I saw the webbing twitch on one of these critters I shined my light directly on it. Through the thin webbing I could actually see a small fish struggling. We continued to watch this octopus for a while, and saw it catch another fish. Perhaps, once these octopuses are out, they continue to hunt throughout the night and only settle down to eat when they return home.

Like his squid counterpart, the male octopus has a hectocotylized arm that he uses to pass sperm sacs to the female — but this is pretty much the only resemblance between squid and octopus mating. For one thing, rather than taking only a few seconds, octopus matings can take an hour or more.

The first time I discovered a mating pair of octopuses, what caught my attention was an octopus totally exposed on top of a coral head. I approached it, and got much closer than I expected, yet the octopus seemed very uncomfortable. It was breathing quickly, waves of color were passing over its body, and it extended its body away from me while keeping its arms in position. I couldn't figure out why it didn't just dash off, until I noticed that one of its arms was extended into a crevice beneath it.

I backed away and circled around, and found the entrance to an octopus den. The octopus on top of the coral head was engaged in mating with the octopus in that den, and wanted to complete the mating while at

Octopuses mating. The female is in the pipe, her suitor extending his arm beneath her mantle. Photo by David Ritz.

Charles Lucy strokes Olivia the octopus as she holds onto my hand.
Photo by C. David Batalsky.

the same time it wanted desperately to get away from me! Being the soul
of tact, I left — after taking just a few photographs for posterity.

I learned that day, and from additional observations, that the female
octopus always seems to be home for the mating, and the male stays outside her den.

I've only observed the beginning of an octopus mating one time,
and it turned out that I — unexpectedly, you can imagine! — became a bit
of a participant. I had been working with an octopus I'd called Oliver, ever
since our first meeting when "he" had readily taken my offering of food.
The unusual thing about Oliver was that he was just as interested — if not
more interested — in my hand as he was in the food. He'd grab my hand
and pull it into his den, and this situation developed to the point that he'd
completely leave his den just to grab my hand and haul it home!

His possessiveness created an interesting problem at first: my
departure. An octopus has eight arms, and each arm has over two hundred
suckers, and each sucker (in an octopus of this size) requires a pull of
about two to three ounces to break its hold — which would mean that if
Oliver used half his arms to hold my hand and the other half to hold himself in his den, I'd have to break a hold of about a hundred pounds to leave!
The situation was complicated by the fact that I didn't want to hurt Oliver,
so instead of jerking my hand away I always pulled steadily. Even con-
sidering that I could lever myself against Oliver's den, there were several
days when I was sure I'd be spending the rest of my life there.

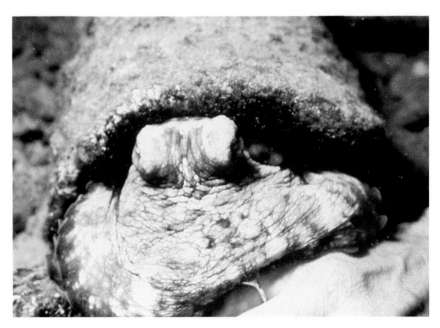

A close-up of Olivia the octopus holding my hand.

However, eventually I'd break loose — usually with a collection of sucker marks on my hand. After the 52-octopus-hickey day, though, Oliver seemed to acknowledge that I was stronger than he was and would release me when I started to pull seriously.

Each day I'd visit Oliver and introduce him to my buddies. Each day he'd grab my hand for reasons I couldn't fathom, and each day he'd eventually release me.

Then, one day, photographer-buddy Tom Downs and I visited Oliver, and saw, nestled against the side of Oliver's pipe-den, another octopus, doing an imitation of a space-in-which-there-is-not-an-octopus. I suddenly realized that this may be a mating situation, so Tom and I settled on the rubble bottom to see what would happen. The new octopus eased next to the opening of Oliver-now-Olivia's den, and stroked the front of her body with one arm. She blushed (and giggled). Then he tried to insert that arm under Olivia's mantle.

Olivia surprised us all: she jilted her octopus suitor! She walked out of her den straight to **my hand**, which she grabbed and towed the ten feet or so back home. She inked as she pulled my hand inside the den, and I think that she was hoping to discourage the blue-and-orange-bubbly-thing (my body) that always accompanied the pentapus at the end of my arm. That scheme didn't work (as you may have guessed), but Olivia didn't seem to mind as she held my hand and investigated it gently with her beak.

During this activity, her confused suitor stayed near the pipe as if he were trying to figure out what he'd done wrong. After months of working on that problem myself, I've come to the only logical conclusion: **he'd** tickled her a couple of times and tried to get friendly — but **my hand** had been bringing her candy and flowers for months!

Under more normal circumstances, the mated octopus female deposits her fertilized eggs in strands on the ceiling of her den. She then aerates and guards them until they hatch. She doesn't leave the den for the entire time; she doesn't hunt and she won't eat. I have knowingly observed an egg-guarding female octopus only one time, and the clue that she might be guarding eggs was that she absolutely refused all food I offered her. After a great deal of contortion, I was able to see beyond her into the den and recognize the unexpectedly small eggs — about the size of rice grains — in strands hanging from the ceiling.

In many species, the female octopus dies when her eggs hatch. Some male octopuses and squid also die after mating.

An interesting aside is that scientists have discovered a gland in the female octopus, which, when removed, seems to prevent her from languishing and dying after her eggs hatch. Instead, she begins eating again and even grows in size. There is hope that these studies can be related to senility in human beings.

CRUSTACEANS
(Phylum Arthropoda, Class Crustacea)
An Introduction to Crabs, Lobsters, Shrimp and Others

The Arthropods (from the Greek, meaning *jointed foot*) are the largest phylum, in terms of both numbers of species and numbers of individuals (since this group contains the insects, and we all know how many of them there are!).

Underwater we find the insects' more tolerable relatives, the crustaceans. Named for their "crusty" shells, crustaceans are primarily marine animals that pass through a planktonic larval stage, breathe with gills, and usually have separate sexes. More than 25,000 species of crustaceans exist, so it's only to be expected that some of them (in this case, the barnacles) won't follow all the rules. . .

Barnacles

Those who explore the marine world around piers and jetties — and from their own boats — are likely to run (literally, as well as figuratively) into barnacles once in a while. I was amazed when I first learned that these hard-shelled critters were crustaceans rather than mollusks.

And yet, when you look closely at a feeding barnacle, you can see little feathery feet — most unmollusk-like — sticking out. The naturalist Louis Agassiz described a barnacle as a "shrimp-like animal standing on its head in a limestone house kicking food into its mouth." The barnacle breathes with gills on its legs.

Barnacles are unique in that they're the only crustaceans that are not free-moving: once they're attached, they're attached for life (much to the chagrin of boat owners). In fact, the glue of some species of barnacles has been studied for use in dentistry!

Barnacles are also unusual in that instead of having separate sexes, they are hermaphroditic. When ready to mate, a barnacle simply uncoils its penis and inserts it into a neighbor (thus the advantage of dense colonies of barnacles). The fertilized eggs are kept within the shell of the adult and released as larvae — up to 13,000 from one animal at a time.

After various larval stages and molts, the young barnacle settles to a substrate, using chemical and tactile sensors to find a hospitable environment — one which is already inhabited by others of its species is especially nice — and begins its adult life by secreting cement to create the plates of its shell and to attach itself to the substrate.

Although barnacles are most frequently seen on jetties and boat bottoms, some species live on whales, turtles, and other marine animals. At best, these hitchhikers must be annoying — any ship owner will testify to the lack of efficiency in the water of a barnacle-encrusted hull, so the same must be true of a barnacle-encrusted whale — at worst, they can be life-threatening. Norine Rouse, who has many turtle friends in the waters off Palm Beach, carries a special little knife that she uses to pry barnacles off any turtle which will tolerate the treatment.

Copepods and Isopods

Copepods are not particularly noted by marine *Homo sapiens,* except when we're plagued by the planktonic "bugs" that surround our lights on some night dives.

On the other hand, isopods, at least the parasitic ones, are often seen by divers. They look like little (one inch long or so) sand fleas, attached firmly around the head areas of various fish. I asked a biologist once if they killed the fish, and he replied "It's a pretty poor parasite that kills its host." Which wasn't much of an answer, but was all I could get; the effect of the parasitic isopod on its host isn't known. One effect I did notice — on a casual basis — was that fish who carry these isopods come in to take food more readily than others of the same species without isopods.

Patrick Colin (*Caribbean Reef Invertebrates and Plants*) made an interesting comment about parasitic isopods that's consistent with my own observations: the species of fish that the isopods live on varies from island to island. For example, in Grand Cayman the squirrelfish and bigeye snappers host the parasitic isopod, in San Salvador the coneys, on Bonaire the creole fish. (Discovering why isopods live on their various hosts in different locations would be an interesting area of study, especially for someone who wants to travel throughout the Caribbean.)

Mantis Shrimp

Although mantis shrimp aren't common animals, one of my most fascinating underwater friends was a mantis shrimp. I write this section with thanks to Godzilla.

Mantis shrimp come in all sizes, but Godzilla, as you may have guessed from his name, is a member of the largest species, growing in size to a foot or more. That shrimp was no shrimp!

My relationship with Godzilla began with my finding his burrow: a hole about four inches in diameter in a sand bottom, completely smoothed around on the inside, and apparently empty. I was curious about what inhabited the burrow, and squeezed a piece of fish at the entrance hoping that fish smell would attract the occupant. This went on for days, and finally Godzilla popped out. My dive log for that day (February 8, 1984) reads: "Those claws are incredible!"

Instead of the pinching-type claw of true shrimp and lobsters, mantis shrimp have a three-segmented arm called a *raptorial claw.* Each segment of Godzilla's arms was about three inches long.

As Godzilla became used to my offering him food, he'd lunge out of his burrow, extending his three-segmented arms and grabbing the fish with the last two segments (like a person holding something between his forearm and upper arm). Often he'd jerk the fish right out of my hand — then he'd disappear into his burrow, not to be seen again that day. More usually I'd keep my grip on the fish, and as Godzilla became comfortable he would stay, his lower body still in the burrow, his claws (not the large raptorial ones, but smaller ones around his mouth) pulling small pieces off the fish and delivering them to his mouth. He'd use his big claws only to keep his balance.

At the end of each interaction I'd give Godzilla a piece of "free" fish — that is, I didn't control it and he could retreat into his burrow with it, which is what he did for the first few months. As he became more comfortable, he began to eat the free food right at his burrow entrance. At

A mantis shrimp looks out of its burrow.

Godzilla the mantis shrimp munches away at the fish I offer.

those times it seemed to me that he was thinking, "Gosh, it's not very often that a fellow has the chance to observe the behavior of marine *Homo sapiens* and eat lunch at the same time!" A real people-watcher, my friend Godzilla.

I could find Godzilla's burrow at night only by knowing precisely where to look. The reason for the difficulty was that at night only his stalked, faceted, sand-colored eyes and his sand-colored antennae protruded from his doorway; the rest of the opening was completely closed with sand. I believe he hunted this way, waiting for an appropriately-sized prey fish to swim along close to his hidden position — whereupon he'd explode out of the burrow, extend his long arms, capture the prey with them, and back into the burrow for a leisurely meal.

We had a storm on Bonaire in November of 1984, and for two weeks afterwards I searched for Godzilla in vain. Finally I spotted a burrow entrance near Godzilla's old one, and when I waved some fish in front of it out popped my old friend. I think his burrow was caved in by the wave action of the storm and it took him that long to dig out! His new burrow wasn't as well-made as the old one, though, and one day I found it collapsed and Godzilla gone. Perhaps he moved to a location where his construction work would be less vulnerable to storms.

When remembering my interactions with Godzilla, I'm most impressed by the fact that this critter, a "mere" shrimp, was capable of learning, of changing his behavior, and of understanding that human beings were not huge predators.

MORE CRUSTACEANS

The Decapods

The Decapods, or "ten-footed" crustaceans, include the common crustaceans: shrimp, lobsters, and crabs.

The head and thorax of a decapod are joined together and covered with a shell called the *carapace* (which contains most of the animal's organs). At the front of the carapace is a pair of stalked eyes and the animal's antennae; at the mouth is a pair of appendages called *maxillipeds* (jaw feet).

The five pairs of legs on the carapace give the decapods their name. The first two or three pairs of legs end in claws and are called *chelipeds* (claw feet); so in addition to the appendages we normally call claws in these animals, there is another pair (or two) of legs ending in less noticeable claws.

Behind the carapace is the segmented abdomen of the animal, quite noticeable in shrimp and lobsters but small and turned under in the crabs. Attached to the abdomen are appendages called the *swimmerets,* which only the shrimp use for swimming. The swimmerets are also involved in decapod reproduction.

Lobsters and shrimp use their long abdomens for fast locomotion; they flip their tails beneath their bodies and zoom along backwards.

Shrimp

The shrimp are the decapods most accessible to divers. As a new diver, the first shrimp I noticed lived in anemones. The most commonly seen of these in the Caribbean is the ghost cleaner shrimp (*Periclimenes pedersoni*), a relatively small shrimp that gets its name from the fact that its shell and body are almost perfectly clear, with just a hint of purplish-blue coloration. Its antennae, like those of almost all cleaner shrimp, are white.

Ghost cleaner shrimp live mainly in and around the ringed anemone

One of the most common and most brightly-colored Caribbean crustaceans is the banded coral shrimp.

(*Bartholomea annulata*), and often several of them live in a single anemone. When they want to clean, the shrimp wave their antennae; small fish, such as harlequin bass and blue tangs, accept the invitation and rest on the bottom near the anemone. As soon as their client arrives, the little shrimp swarm over its body, removing dead scales and parasites from the fish.

Another shrimp that also lives with the ringed anemone but is more hidden than the ghost cleaner shrimp is the snapping shrimp (*Alpheus* sp.), also called pistol shrimp. Their presence in an anemone can only be detected by two pairs of barber-pole striped antennae — or, if an unwary diver sticks a hand too close to the anemone, a surprisingly vigorous *thunk!* on a finger. If a piece of fish is held near the antennae, eventually these shy shrimp will show enough of themselves to grab the food and drag it back under the anemone. Snapping shrimp look like tiny (just over an inch in size) American lobsters. They use their larger claw to *thunk!* against any intruders to their anemones.

A scientist once removed the snapping shrimp from all the ringed anemones in a given area; a short time later he returned to find many of the anemones missing, but the anemones in his control group (with their snapping shrimp still in residence) were all there. So it appears that the snapping shrimp help protect their anemones from predators as well as curious divers.

Even without something to *thunk!* against, snapping shrimp make a snapping noise with their claws that strikes terror into the heart of any aquarist who is keeping them, since the noise sounds exactly as if the aquarium glass is cracking.

One summer I worked with a group of scientists who were attach-

A nocturnal Caribbean shrimp resting in a gorgonian. 1:3.

ing beepers to sharks to track their movements; out in the Bahamas Bank
in the middle of the night, tracking sharks with a receiver dunked into the
water, we were often drawn off-course by snapping shrimp noises that
sounded exactly like the beepers!

A relative of the ghost cleaner shrimp, the spotted cleaner shrimp
(*Periclimenes yucatanicus*), usually lives in the common Caribbean anemone (*Condylactis gigantea*). This shrimp grows bigger than its cousin and is
named for its beautiful purple spots. The spotted cleaner shrimp is much
less likely to clean than the ghost cleaner shrimp, and despite the fact that
it waves its antennae just as invitingly, I have never seen the invitation
accepted. I wonder if the spotted cleaner shrimp is just trying to lure
unwary clients into its anemone partner, although I've never seen that
happen, either.

The "sticky," or "sun" anemone (*Stoichactis helianthus*), hosts still
another *Periclimenes* shrimp, a mystery creature that's only mentioned in
one of my books (Colin), and even there its species is not identified. This
shrimp is clearly not a cleaner of fish, since it doesn't have white antennae
and never acts as if it wants to clean. It's perfectly camouflaged within the
light brown tentacles of the sticky anemone — much of the time I can
only find it by looking for the tentacles it bends as it walks along the surface of the sticky anemone.

While each of these *Periclimenes* shrimps is found on a specific
anemone, there is another little shrimp who finds shelter in many different species of anemones (including the *Periclimenes* shrimp's anemones):
Thor amboinensis. Since this shrimp has no common name, I call him
Thor, and wonder if he was named after a thunder god as a joke. Thor is
small even for a shrimp, growing only to about half an inch in length. Thor
shrimp have short brown antennae and carry their tails high up as they

jump around the stalks or within the tentacles of their anemones. Several Thor shrimps usually live with each anemone, and they're much more active than the other anemone shrimps.

One of the favorite shrimps of new divers, and frustrating models for underwater photographers, is the banded coral shrimp (*Stenopus hispidus*). The first time I saw a banded coral shrimp was when I was a very new diver, diving with an experienced buddy near Miami. My buddy looked into a tube sponge and carefully squeezed it upwards until a banded coral shrimp popped out. For months afterwards I looked into every sponge I passed, hoping to find my own banded coral shrimp. In all the thousands of dives I've had since then, I've seen plenty of animals living in sponge cavities, but never another banded coral shrimp!

Eventually I learned to find them by looking under ledges and in crevices for their long white antennae. Banded coral shrimp have long claws and are red and white (but may look black and white, depending on the light level where they're seen). Their colors and delicate appearance are what attracts underwater photographers; their delicacy — their habit of detaching a claw or two when they're stressed — is what frustrates these same photographers.

Banded coral shrimp almost always live in pairs; the one with a greenish coloration on its underside is the female, and the greenish stuff is the eggs she's carrying.

Banded coral shrimp have the long white antennae that are the guild sign of cleaner shrimp, but they seem to be only occasional cleaners. Their favorite daytime clients are moray eels. On a few night dives I've seen fish poised near banded coral shrimp, so perhaps they travel around then and clean sleeping fish....

A close relative of the banded coral shrimp is the yellow and orange-red striped golden coral shrimp (*Stenopus scutellatus*). It's less common than the banded coral shrimp (or at least more difficult to find), and lives in deep caves or under rocks. An endearing characteristic of this shrimp is two bright-red mouth parts that always turn around in circles; since these bright spots are more noticeable than the shrimp's eyes, it appears as if the critter has constantly-spinning red eyes.

My favorite shrimp is the scarlet lady (*Lysmata grabhami*), also called the red-backed cleaner shrimp, a pretty shrimp with a white stripe down its red back. Scarlet ladies reach up to two inches in size, they tend to live in pairs, they have the characteristic white antennae of cleaner shrimp, and there is no doubt that they are cleaners. They're just as willing to leap upon the hand of a diver as they are to clean a fish, and a pair of them once cleaned my hand so vigorously that they reopened a cut on my finger!

A sharp-eyed reef watcher can find many different kinds of shrimp. Tiny shrimp called *mysids* live in swarms around the tentacles of some anemones, including the ringed anemone. Flattened shrimp live on whip corals. Pairs of glowing red dots seen in the light on night dives are the eyes of reef shrimp — as the diver approaches, the shrimp jumps out of sight. Purple shrimp live in pairs on Spanish dancer nudibranchs.

Shrimps live with many echinoderms, too: a tiny flattened shrimp lives on the spines of some urchins, and many sea stars are accompanied by a pair of shrimp whose coloration perfectly matches the stars' bodies. A study of the arms of crinoids will often reveal a small shrimp or two, perfectly colored to match the crinoid: orange shrimp in orange crinoids, black shrimp in black crinoids, black-and-white shrimp in black-and-white crinoids. On night dives the apparently black coloration of the crinoid shrimp shows its true red color, and the shrimp are easy to see against their black backgrounds. The West Indian sea egg, a short-spined, black-and-white urchin, often hosts one of the best camouflaged shrimp I've ever been clever enough to find. This little shrimp is about the same length as the sea egg's spines, and holds itself upright between the spines, where its black-and-white coloration camouflages it to look like two urchin spines.

Lobsters

A lobster is constructed pretty much like a huge shrimp with an extra-sturdy exoskeleton. Like shrimp, all lobsters have claws, although some have more obvious claws than others. Included in the lobster group are the American lobster (*Homarus americanus*), the one most of us think of when we hear the word "lobster," as well as the Caribbean's spiny lobster (*Panulirus argus*). Many northern divers mistakenly believe that just because the spiny lobster doesn't have huge claws, it has no defense mechanisms — but anyone who has tried to catch one of these critters knows how false that assumption is!

The spiny lobster is, as its name implies, covered with spines all over its body (including its antennae). These spines are such an efficient defense that holding a struggling spiny lobster barehanded generally results in shredded fingers.

Getting a spiny lobster to hold in the first place is a difficult task: these critters and their relatives live under coral heads and in crevices in the coral, facing outward. When a lobster hunter reaches into the crevice, his or her glove is immediately halted by the outward-facing spines of the lobster's antennae. Grabbing a lobster by the antennae is totally ineffective, since the antennae simply break off and the lobster retreats into its crevice, snickering.

Breaking the antennae off these animals is not really a laughing matter, though, since their antennae are their first line of defense. I learned this the time I decided to try to feed a large spiny lobster. I held out the food, which to my chagrin attracted the attention of a number of yellowtail snappers as well as that of the lobster. The lobster began munching on the food in my hand, at the same time keeping the yellowtails at bay by using its armored antennae like rapiers. Those yellowtails — experienced

food-snatchers — were kept completely at bay by the expert swordsman-
ship of the lobster.

A type of Caribbean lobster that lives up to the nickname "bug" is the
slipper, or shovelnosed, lobster (*Scyllarides* sp.). This lobster's antennae,
instead of being long, are short and paddle-shaped; it has a set of lovely
purple short secondary antennae and two wide-set tiny eyes. Despite the
fact that slipper lobsters look a lot like large beetles, they're appealing.

Slipper lobsters are nocturnal; the only ones I've ever found in the
daytime have been flattened upside down against the underside of a coral
head or crevice and look like a bump on the rock. At night they wander
around, usually on sand areas, and one time I found one munching on a
clam whose shell it had cracked.

The Caribbean has a few varieties of lobsters with claws, but they're
uncommon. Under Bonaire's Old Pier I was lucky enough to find a pair of
very beautiful clawed lobsters that were six to seven inches long and
orange-brown-red in coloration. The claws were large in relation to the
size of the lobster. On one of the critters (the female) the claws were
smooth and on the other (the male) they seemed hairy. At first the lobsters
were shy, but after only a couple of days of being offered food, the male
began to come out from his crevice, and soon he was fearlessly eating the
food from my hand. If I let go of the food, or if he pulled a piece free, he'd
bring it under the crevice and share it with his mate.

Hermit Crabs

Remember the time you found that lovely empty shell? You snatched up
your prize to carry it home, and suddenly, a few minutes later, it pinched
you! — or at least the hermit crab, who had retreated well inside the shell
as you grabbed it, emerged to protect its home.

The body of a hermit crab only has the "crusty" shell in front; the
animal's abdomen is soft and unprotected — so it must **find** protection,
instead of growing it as other crustaceans do. Usually hermit crabs live in
uninhabited shells, but some have been found in more unlikely homes,
such as discarded tin cans.

Some sand flats in the Caribbean and Bahamas are inhabited by
numbers of tiny white hermit crabs, well under an inch in length including their tiny white shells. If I chased one of these crabs with my finger, it
would zoom over the sand at a surprising speed. One day the fellow I was
chasing, suddenly tiring of carrying all that excess baggage around, leaped
out of his shell and continued his escape unencumbered! Luckily for his
life and my conscience, I was able to cup him in my hand and bring him
close to his shell; soon he backed into it and ambled off. His unusual
defense mechanism didn't seem to be very functional to me, unless it
might fool a predator into gobbling up his shell instead of him!

The biggest species of hermit crab lives in the shells of queen
conchs. These crabs have also been seen eating queen conchs, but it is not

◄ *This hermit crab's body is more colorful than its shell! 1:3.*

yet known if they are capable of actually attacking the conch and then taking over its shell.

One of the most interesting habits of some hermit crabs is to detach a particular type of anemone from its substrate and re-attach it to their shells. The crab gets the protection of the anemone's tentacles; the anemone gets a moving substrate, which increases its chances of acquiring food, plus it may also get leftovers from the crab's meals. The hermit crab/anemone partnership is most common in Pacific oceans, but I have seen it a couple of times in the Caribbean.

True Crabs

Crabs come in varying sizes, from pea crabs well under an inch in length to the giants found in Alaskan and Japanese waters. Crabs live in many different environmental niches, too: from shoreline to reefs, from sand to rock, from warm waters to cold waters.

The crabs are the most advanced of the decapods. They have one pair of small antennae. Their eyes, on movable stalks, can be pulled back under the protection of the carapace. Their ten legs are divided into four pairs of walking legs and one set of claws. Their abdomen, much smaller than its counterpart in shrimp and lobsters, is turned back under the shell.

Their hard exoskeleton is not the crabs' only protection; they rely on

We found this spider crab on a night dive off Little Cayman.

I found this long-legged crab living in a clump of the same type of algae it's camouflaged with, on Bonaire.

additional methods, such as speed or camouflage, to foil their predators.

The less protected an environment a crab lives in, the quicker it seems to be. My friend's dog used to spend hours poised on a dock in the Florida Keys, leaping into the water after crabs that always eluded her grasp. The same crabs were perfectly willing to pinch the bottom of anyone sitting on the sand!

The most familiar crab to the general public is probably the blue crab, but to Caribbean divers it is surely the arrow crab (*Stenorhynchus seticornis*). This creature, with its tiny arrowhead-shaped body and disproportionately long legs and claws, lives under crevices or near the protection of anemones and long-spined urchins. It's fairly easy to capture, but not above giving its captors a nice pinch for their troubles.

Most crabs are active at night rather than in the daytime; a reef area that seems totally crabless even after many daylight dives exposes its crab population after dark.

Caribbean divers who see a large crab grazing on algae at night have almost always found a spider, or mithrax, crab (*Mithrax spinosissimus*), one of the Caribbean's largest crabs; the carapace of this animal can grow to several inches across. It's easy to tell mithrax males from females since

This sponge crab's sponge is too small to fool anyone!

the males have much larger claws. Once you have one of each lined up for you, the difference is obvious.

A small relative of the spider crab, another mithrax crab (*Mithrax cinctimanus*), spends its days under the protecting tentacles of anemones. Like its large cousin it also grazes at night.

Despite their hard shells and efficient claws, stone crabs (*Menippe mercenaria*) spend much of their time in burrows in sand or mud bottoms. Their relatives, the coral crabs (*Carpilius corallinus*), live in crevices in the coral reef and are occasionally found by divers, even in the daytime; their red and cream coloration gives them away. I decided to feed a coral crab once, and held out a piece of fish expecting the crab to reach out gently with its big claws and take the fish from my hands. Instead — much to my surprise — the crab leaped onto my hand, held itself in position with its big claws, and began munching away at the food as it rested on my hand! I was amazed that it made no effort to take the food away from me and bring it back into its crevice, as lobsters always do, and I was also very relieved that it made absolutely no effort to pinch me with its claws.

At least two kinds of crabs take advantage of the fact that their predators find sponges distasteful, and camouflage themselves with sponges. The decorator crab of the family Majidae plucks small pieces of sponge and "plants" them on its shell, on its legs as well as its carapace. The sponges are thought to be attached to the shell of the crab by small hooked hairs at first, but soon they begin to actually grow on their new substrates. Some types of decorator crabs prefer using only one color sponge, while others may have several differently colored sponges grow-ing on them. The more the sponges grow, the more they conceal the crab-like shape of their host, and the differently-colored sponges also break up the outline of their crabs.

Like most crabs the decorator crabs are active at night. Even though I know what they look like, I only very rarely can find one in the daytime. Their predators would have the same difficulty in spotting them that I do — but even if a predator did locate one of these crabs, it would be unlikely to gobble down the tasty morsel because of the not-so-tasty sponges. The decorator crabs have an efficient multi-level system of protection from predators.

Another crab that takes advantage of the protection of sponges is the sponge crab (family Dromiidae), a stocky, dull-colored crab with maroon tips on its claws and legs. Rather than planting sponges on its shell as the decorator crabs do, the sponge crab carries its protection around with its last pair of legs. How the sponge crab acquires its sponge is a bit of a mystery, but I suspect it uses its claws to clip off a likely piece of sponge. After the crab has carried the sponge for a while, the inner surface of the sponge becomes rounded to fit the crab's carapace perfectly.

Sponge crabs are algae-grazers — no sponge crab has ever respond-ed favorably to my offers of a scrap of fish — but they live on sponges. They're well-adapted to this life since the pointed tips of their four pairs of walking legs dig into the sponge and make them difficult to remove. Their last pair of legs, the pair that holds the sponge in place, can't be used for walking because these legs are turned upwards over the carapace of the crab, specially adapted for carrying the sponge umbrella.

A sponge crab whose sponge is trapped (for example, by a diver try-ing to get a closer look), may abandon its sponge. Since the crab's legs are constructed for walking, not swimming, it sinks to the bottom and scurries away. If the crab is reunited with its sponge (say, by the guilt-ridden diver), it may choose to take the sponge back. This process looks like a person putting on a jacket: the crab puts one leg into the sponge, swings the sponge around, places the second leg, and shrugs the sponge into posi-tion. Feeling safe again, the crab ambles off.

Finding a sponge crab in the daytime can be quite a trick, since sponge crabs prefer spending their days on the same sponges from which they've taken their own sponge. They squeeze their bodies into the broken part of the sponge, fit the sponge they're carrying over them like the lid on a cookie jar, and snooze the day away. No crab is visible, only a sponge with a cut across it.

I've seen two-inch sponge crabs carrying six-inch sponges and six-inch sponge crabs carrying two-inch sponges, but usually the crab and its protecting sponge are well-matched in size.

Box crabs (*Calappa* sp.) don't need to hide beneath sponges; they seem to materialize and dematerialize at will. They're named for their boxy appearance; their claws and legs fit up against their bodies in perfectly-shaped indentations in their shells. They primarily eat gastropod mollusks, crushing the shells with their strong claws, but they often find me when they're attracted by the scent of food I'm offering another animal. Suddenly they just appear out of the sand. When they're ready to disappear, they rest on the sand, push sand away from them with their claws, cuddle down into the resulting depression, and continue the process until only their stalked eyes protrude from the sand. This is the same system other crabs that hide under the sand use, but the box crab is considerate enough to do things slowly, so we can see what's happening.

Crabs best-adapted to swimming, like the blue crab, have flattened, elongated bodies that often come to a point on each side, making them hydrodynamically efficient as they swim sideways through the water. They use the flattened last segments of their last pair of legs to scull themselves along. Anyone who's tried to catch a swimming crab knows how quickly they can move.

The most common swimming crab in Bonaire's waters is *Portunus sebae*, a light-brown crab with one dark-brown spot on each side of its carapace. When cornered, this crab spreads its claws out wide in a defensive gesture and is quite willing to grab an extended finger in a claw — an interaction in which the owner of the finger is usually not so willing to participate. If, instead of a finger, a piece of hot dog is extended, the crab grabs that, too — then suddenly realizes it has food and retreats, munching happily away.

This swimming crab is often found in pairs, with the smaller (female) crab cuddled between the claws of the larger (male); even when they move around or feed, they do so in tandem. One may conclude that this behavior is related to mating, and one would be correct.

EVEN MORE ABOUT CRUSTACEANS

Molting

Their hard shell gives crustaceans excellent protection from predators, but makes mating, or even growing, difficult. Thus, crustaceans must molt, that is, shed their hard shell and produce a new one.

I've only seen crabs in the process of molting two times, and on this dive I was lucky enough to have a buddy who could immortalize the event on film. The old shell is facing left; the soft-shelled crab's flexibility is shown by the curve in the point at the top of its shell. Photo by Mark Krakowski.

The molting process is controlled by hormones. When the animal is ready to molt, its body begins to swell (in some animals the swelling is helped by the critter reabsorbing calcium salts from its shell). The shell then splits. The animal struggles free of its old shell, and soon its "skin" hardens into a larger, bright-and-shiny new shell.

Finding a lobster with a growth-encrusted shell doesn't necessarily mean that you've found an old lobster, but rather that the encrusted one hasn't molted in a while. Younger crustaceans molt more often than they do when they've reached adult size.

Most divers have been fooled by a lobster molt at least once: "Wow!" they think, "Look at that huge lobster, right out in the open!" They rush over to pick it up, and it's limp in their hands — a molt. The reason we all get fooled is because the molt is the entire surface covering of the animal, right down to the covering of the eyeballs.

There are two differences between a molt and a dead crustacean: a dead animal will usually have some meat left in the shell (and a billion small fish hanging around, trying to get at that meat); a molt is clean inside, and split along the line between the carapace and the abdomen. This split is easier to see with crab and lobster molts, because of their larger size and thicker shells, than it is with shrimp molts.

During molting, and before the shell hardens after a molt, are the times a crustacean is most vulnerable to predators. It keeps a very low profile then, and usually molts under cover of darkness. I was lucky enough to see an arrow crab molt once, and it's too bad we can't watch molting more often because it's very entertaining. Imagine a ten-limbed diver trying to take off a tight wet suit, and you'll understand what I mean. That poor arrow crab had gotten out of his old carapace when we found him, but it took him another twenty minutes to free his legs!

Reproduction in Crustaceans

Most crustaceans can't mate until just after the female molts. The male crab molts before the mating season. When the female is nearing readiness to mate, she is believed to release a scent to attract the male. He stays with her, protecting her (and preventing her from leaving!) until she molts. After the female molts, when her shell is still soft, the male is able to insert his swimmerets (modified appendages on his abdomen, which in crabs have nothing to do with swimming) into her genital opening and pass sperm to her. He then continues to protect her until her new shell is completely hardened.

Later, as the female releases her eggs, they are fertilized by the sperm she has stored. She attaches them to her swimmerets, which are greater in number than the male's and "hairy," providing plenty of surface for the sticky eggs to adhere to. The egg clusters of crabs are spongy-looking, and their light color becomes darker as the eggs develop. After a few weeks the

After the smaller of these crabs, the female, molts, the pair will mate. The male will stay with her until her new shell hardens.

eggs hatch and the tiny shrimp-like larvae enter the plankton layer, where they molt several times until they attain a larval stage that looks more like a crab than a shrimp. This *megalops* larvae settles out of the plankton layer to the bottom, and at its next molt becomes a juvenile crab.

Of the lobsters, the American lobster's reproductive habits are best known. The female lobster, like the female crab, must also be soft-shelled when she accepts sperm, and her eggs are also fertilized as they pass from her genital opening and before they attach to her swimmerets. The incubation period of lobster eggs, during which the female aerates and protects the egg cluster, is nine to twelve months. Then the eggs hatch and the larval lobsters enter the plankton layer.

Of all crustaceans, the mantis shrimp female probably takes the most trouble over her eggs. Once they are released and fertilized, they don't stick to her body. She makes a (ping-pong-sized) ball out of them and tends them with her three-segmented arms, making sure they stay clean and aerated. During the weeks it takes for her eggs to hatch, she doesn't eat.

Food Versus Fascination

People are most familiar with the crustacean as an animal they meet on a dinner plate (or one they're trying to invite for dinner), but these fascinating animals are well worth interacting with instead. High on my list of Things To Do (but not as high as Dive in Cold Water) is Feed an American

Lobster. Yeah, I know all about the Crushing claw and the Ripping claw, and I expect to start out cautiously by offering the food on a lo-o-o-ong stick. I also know about all those Lobster-Hunting divers, so obviously the site to begin this activity wouldn't be Site Number One on the Top Ten List of Northern Dive Sites. Still, lobsters are willing to eat just about everything, and I think it would be fun and a great compliment to have a lobster friend walk out to greet me and climb trustingly on my arm as he or she munched away at some hors d'oeuvres.

ECHINODERMS
(Phylum Echinodermata)
Sea Stars, Urchins, Sea Cucumbers and Crinoids

The echinoderms (from the Greek *echinos,* meaning hedgehog or spiny, and *derma,* meaning skin) consist of over 5000 species of marine animals that inhabit waters from the shore zone — some urchins dig out crevices for themselves in rock that's actually out of the water during low tides — to several hundred feet of depth. They inhabit waters of all the temperatures and locations that a diver is likely to explore.

A unique characteristic of the echinoderm group is their water vascular system, an internal hydraulic system of canals that end in tube feet; the tube feet are extended or retracted as the animal's muscles adjust water pressure within the canals. Many echinoderms use the water vascular system for breathing. In addition, tube feet help them stick to hard surfaces, dig through sand, or clean sediment from their bodies.

The reproductive system of echinoderms is basically the same as that of the sponges and some of the mollusks. Their sexes are separate (in almost all echinoderms); external sexual characteristics aren't necessary because sperm and eggs are simply released into the sea. I've observed the spawning of the West Indian sea egg (*Tripneustes esculentus*), a black-bodied urchin with short, fairly blunt white spines. A few West Indian sea eggs clinging to the dead bases of some staghorn coral were releasing a thick white fluid (that looked a lot like Elmer's®glue). As I watched, other urchins also secreted their gametes.

The same thing happens with heart urchins, except that it's more mysterious: since heart urchins are generally buried, their milky reproductive fluid seems to be squirting out of the sand. A closer look reveals the urchin beneath each tiny volcano.

The fertilized eggs of echinoderms drift in the plankton layer as they develop into larvae; they eventually settle to the bottom as miniature adults.

Larval echinoderms are bilaterally symmetrical — that is, if you drew an imaginary line down their centers they'd be the same on both sides (like people). Once they reach adulthood, however, most echinoderms appear to be radially symmetrical with no head and a body with five rays, or multiples of five rays, extending away from its center.

The echinoderms are divided into five classes: the sea cucumbers, the crinoids, the brittlestars and basketstars, the sea stars, and the urchins.

Sea Cucumbers *(Class Holothuroidea)*

The sea cucumbers are the echinoderms that look least like they should be members of this group. Since they really do look like cucumbers, the echinoderm's radial symmetry isn't obvious — unless you're looking at one of the sea cukes that has five arms protruding from its mouth to gather food. That type gives away the fact that their radial symmetry works from end to end.

Most tropical-water sea cucumbers live on (or under) sandy-type bottoms, where they remove detritus from the sand as they pass it through their bodies. According to one estimate, a Caribbean sea cucumber passes five hundred to a thousand **tons** of sand through its body each year! Some sea cukes leave behind them characteristic sausage-shaped deposits of filtered sand called fecal casts. When I first started diving, I carefully took a series of photos of a pile of fecal casts I'd found on the bottom near a moray eel, thinking it had something to do with the eel. As I described my

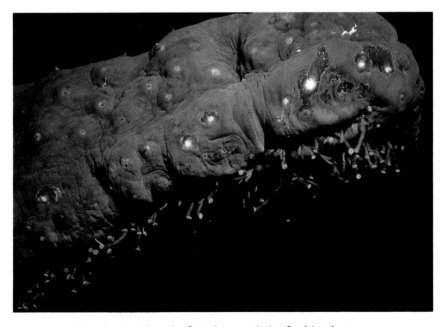

A sea cucumber, showing the tube feet characteristic of echinoderms.

The lion's paw sea cucumber has bumpy skin that mimics spines.

"mystery sand formation" to an audience of (thank goodness!) nondivers, I suddenly realized what we were seeing. I mumbled the rest of the description and sentenced the slides to anonymity in my archives.

Still, those undersea sand sausages do give us some information: we can tell that a sea cucumber's in the area, we can tell if it's been in that spot for a relatively long or short time, and we can tell one end of it from the other. Another way to determine the head from the anus of a detritus-eating sea cucumber is this: the mouth is aimed downward and the anus straight out.

One sea cucumber that's different from most members of its class in the Caribbean is the one I call the sticky sea cucumber (*Euapta lappa*) because it sticks to anything that touches it. Its thin skin is almost clear and it hasn't got the tube feet its relatives use for locomotion. It would be difficult to identify as a sea cucumber at all, except for the five arms it uses to brush the bottom for food.

Like the sticky sea cucumber, cold water sea cucumbers — such as the ones off California — are recognizable by the telltale five arms that distinguish them from filter-feeding worms or anemones. The bodies of these colorful sea cucumbers are usually smaller than their Caribbean relatives, and are hidden in rock crevices; only their feeding activity is visible as they move one mucus-coated arm after another into their mouths to remove the attached food particles.

Although they're echinoderms, sea cucumbers don't have skin that is literally spiny. Some of them do have fake spines on their bodies. The Caribbean lion's paw sea cucumber (also called the elephant's trunk sea

I found one species of sea cucumber in Fiji that would defend itself this way quite readily. Close-up.

cucumber, *Holothuria thomasae*), has fleshy spines up to half an inch long that protect it by discouraging overzealous divers from touching the "spines"; perhaps this system is equally effective with predators.

The lion's paw sea cucumber feeds most actively at night, stretching its body to five feet long or more. Its mouth end extends and the other end of its body is hidden under a coral head. Anyone who tries to pull this sea cucumber free will find that it's a lot more difficult than one might think — how does the critter stay put so tenaciously?

I discovered the answer the night I found a lion's paw sea cucumber situated in a crevice (rather than under a ledge), so I could see the entire animal. I tickled the head end, and realized that as the animal retracted, it didn't expel the water it was filled with, but rather pushed that water into its rear end. That's how they hold so tightly: their bodies are so swollen with water they're stuck in their crevices!

Many sea cucumbers use evisceration as a defense mechanism. When threatened, the sea cucumber discharges its internal organs (which it will later regenerate) out its anus; with luck, the predator will be content to feed on the internal organs and leave the sea cucumber itself alone.

Other sea cucumbers eject from their anal cavities not their internal organs, but sticky tubes (in some cases poisonous) called the tubules of Cuvier. The yucky mess discourages predators, and can even entangle small animals that may have been disturbing the sea cucumber.

The muscles that a sea cucumber uses to eject its internal organs or its tubules of Cuvier are probably the very same muscles the animal uses to breathe, since sea cucumbers draw water in through their anal cavities to respiratory trees within their bodies. No wonder Kaplan calls the Holothurians a "bizarre echinoderm class" (*A Field Guide to Coral Reefs*, p. 193)!

Crinoids *(Class Crinoidea)*

There are 5000 identified fossil forms of crinoids but only 620 currently known living species! Reef crinoids are sometimes called sea lilies, but technically the sea lilies are a deep water (300 feet and more) group of stalked crinoids.

The crinoid's body, called the *calyx*, is relatively small (one to two inches in diameter), and contains the animal's mouth, digestive tract, and other organs. The crinoid's mouth is at the top of its body (toward the surface of the water) and is surrounded by up to 200 arms that are twelve inches or more long. From each arm, rows of mini-arms called *pinnules* project sideways, and on each pinnule are tiny tube feet with even tinier mucus-secreting projections called *papillae*.

A food particle that drifts within the net of the crinoid's arms is trapped in the mucus and moved down a groove in the pinnule to a groove in the arm and into the crinoid's mouth. This system is similar to that of the filter-feeding worms, but on a larger scale.

On the underside of the crinoid's body are the *cirri*, stiff short

The Coral Sea had more crinoids than any other place I've dived; also, each individual had many more arms than its Caribbean counterparts.

A close-up of one arm of a Red Sea crinoid.

A diver examines a crinoid colored very much like a lionfish.

"legs" with which the crinoid holds on to the substrate.

Crinoids are among the most colorful inhabitants of the tropical reef. There are red crinoids and yellow ones and black ones and maroon ones; there are black-and-white crinoids and red-and-white crinoids, and yellowish-green crinoids and. . . .

Crinoids are as interesting to touch as they are lovely to look at. If you run your finger gently down the arm of a crinoid, you can feel the tube feet grabbing onto your skin; rather than sticking like the tube feet of an urchin or sea star, however, these tiny feet feel just like one piece of Velcro™ grabbing another. The crinoid's grip is as strong as Velcro,™ too, so we have to be careful that we don't accidentally take a piece of crinoid away with us after we touch it.

Some crinoids perch out in the open with their bodies and arms exposed; others keep all of their bodies and most of their arms hidden in crevices and filter the water with only the tips of their arms. Some crinoids stay in the same place all the time, while others move into the open only at night or only in the daytime. Most crinoids move slowly by walking along on their cirri, but some, especially in the Indo-Pacific, are able to actually "swim" with coordinated movements of their arms.

Like most echinoderms, crinoids provide homes for a number of small animals. In the Caribbean, almost every crinoid has one or more little shrimp living within (and probably stealing food from) its arms. In Indo-Pacific waters, gobies and other small fish, also perfectly-camouflaged, live within the arms of crinoids.

In the Red Sea there is a common crinoid whose coloration almost perfectly matches that of a local lionfish. The question is, does the crinoid mimic the venomous lionfish for protection (not that there are many known predators of crinoids), or does the lionfish mimic the crinoid to camouflage itself from its small-fish prey (who don't seem to spend time hanging around crinoids anyway), or is the resemblance a coincidence?

Brittle Stars and Basket Stars (Class *Ophiuroidea*)

The central disc of a brittle star is similar to a crinoid's body, but the brittle star's five arms usually radiate sideways rather than upward. Each of the brittle star's arms has a skeleton of calcareous plates, and many also have surprisingly strong and sharp hair-like spines.

Brittle stars can move more quickly than most other echinoderms. They lead off with (any) two neighboring arms, "row" along with the next two arms, and sort of drag the fifth arm. The star's body is lifted off the bottom by its arms as it moves, and the spines on each arm help the star keep traction.

This quick locomotion is demonstrated by brittle stars every time

A brittlestar on a sponge, Belize.

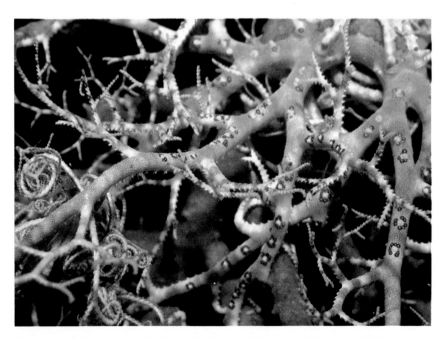

The arms of a basketstar, 1:1. The rolled up tendril on the left has probably captured some planktonic matter.

they're exposed; they stay hidden in the daytime, so turning over a rock is likely to reveal several hairy-armed brittle stars scurrying back to darkness. (Divers should be cautious in rock-turning, since if this activity is observed by a puddingwife, say, or a Spanish hogfish, the brittle stars are likely to be eaten before they can return to safety, regardless of how fast they move!)

Some brittle stars live beneath the sand, and are only revealed by their five sand-colored arms, reaching well above the surface of the sand to filter food.

Besides living under sand and rocks and within crevices in coral heads, brittle stars also inhabit spaces in and around sponges and are well-represented in the Caribbean's tube sponge "condominiums." During daylight hours, the condos are inhabited by numerous brittle stars companionably avoiding light — and predation. At night, the brittle stars desert the inside of the sponge's tubes and decorate the outside of the tubes, while their diurnal apartments are tenanted by sleeping fish. At the same time, they are able to filter food from the currents of water passing around the sponges. They probably don't affect the sponge's food-filtering, since sponges feed on smaller planktonic materials than do other marine filter-feeders.

The speed, retiring habits, and armor of brittle stars aren't the only ways they avoid predation. They have one more trick: disengaging a captured limb (the term for this is *autotomy*). If a predator has grabbed a brittle star by an arm, the star releases that arm (like a lizard may release the

At night the basketstar expands to feed.

captured end of its tail). If it reaches safety, the brittle star (like the lizard) will eventually regenerate its sacrificed body part.

The brittle star's closest relative and companion ophiuroid is the basket star, whose body is tiny compared to its mass of arms. Like its relative, the basket star has five arms attached to its body. **Un**like its relative's, the five arms of the basket star immediately branch again and again. Basket stars are in the logically-named family Gorgonocephalidae (Gorgon's head).

Basket stars live in areas of relatively strong currents, and spend their daylight hours curled up in a tight ball under coral heads or in crevices. In the Caribbean, some basket stars live almost exclusively with a tall purple gorgonian; the tightly-balled basket star looks like a bird's nest in the tree-like gorgonian. An examination of such a bird's-nest basket star (keeping in mind that too much pawing will unduly stress the basket star) may reveal the existence of a tiny resident shrimp whose mottled pale coloration perfectly matches the coloration of the basket star.

At night basket stars change from introverts to extroverts. Those that spent the day in crevices move onto the tops of coral formations. Basket stars expand their arms into the current, using the resulting lace-doily-type net to catch large planktonic critters such as shrimp. On a night when plankton is readily attracted to your dive light, shine it near a basket star and you'll be able to watch the star's tendrils grab tiny animals and deliver them to its mouth.

MORE ECHINODERMS

Sea Stars *(Class Asteroidea)*

Since "starfish" Are not fish, biologists prefer to call them sea stars. (This accuracy in common names hasn't yet spread to sea cucumbers. . . .)

The sea stars are good examples of echinoderms: they have bilaterally symmetrical larvae that develop into radially symmetrical adults with five equal arms radiating from a center area (rather than a specific body disk like the brittle stars and basket stars). A sea star's mouth is on its oral (underside) surface; its waste products diffuse through its skin or are expelled through its water vascular system.

The water vascular system is more highly developed in the sea stars and urchins than in the other classes of echinoderms. Sea stars are well-known for their ability to force open bivalve mollusks by pulling against their shells with relays of tube feet. Once the shells separate as little as one sixteenth of an inch, the sea star extrudes its stomach into the space and digests the body of the mollusk within its own shells! Grisly but efficient.

All sea stars are predators whose main food source is clams and oysters. This is part of the reason that they're not commonly found in warm-water reef areas. In tropical seas, bivalves (and other mollusks) are most likely to be found around sand flats. On San Salvador the only time we'd see sea stars was when we traveled to the conch beds, which the conch shared with lots of large sea stars called West Indian sea stars (*Oreaster reticulatus*).

A few types of sea stars do live near coral reefs. They are most likely to be found at night when they move along the surface of the sand. In the daytime they bury themselves under the sand.

The most infamous sea star of recent years has been the crown-of-thorns (*Acanthaster planci*), which feeds on stony corals. The population of crown-of-thorns sea stars on Australia's Great Barrier Reef suddenly

A crown-of-thorns starfish.

seemed to increase, and enormous areas of corals were being turned into areas of dead coral skeletons by these voracious predators. Panicky divers and fishermen tried to rid the Barrier Reef of the sea stars, and at least one of their measures was counter-productive: some people were cutting up the stars to kill them. Instead of dying, though, these adaptable echinoderms were simply growing back from the pieces: a single crown-of-thorns that had been cut into, say, five pieces may grow into five separate crowns-of-thorns — which was definitely not the goal.

Ultimately, more reliable methods of reducing the crown-of-thorns population were found (one was injecting the sea stars with formalin, another was suggested by a wit who wishes to remain anonymous: "Inform whichever groups that are poaching from the Great Barrier Reef that the crown-of-thorns sea star is a national resource and is protected from all harvesting. They will suddenly perceive it as valuable and steal hundreds of specimens!").

Once the formalin proved effective, the question of "How can we reduce the crown-of-thorns population?" was replaced with "What caused this population explosion in the first place?" At first it was thought that removal of trident trumpets (a mollusk that is a natural enemy of the crown-of-thorns) had contributed to the sudden increase of the sea stars, but finally it was decided that the population growth was part of a natural cycle. Sort of an anticlimactic ending to a scary story, especially since only time will tell if that answer is correct.

One reason for the panic at the increase in the crown-of-thorns population was the fact that the sea stars were destroying the coral reef; another reason was that crowns-of-thorns are very unpleasant to encounter: they live up to their name by being covered on their top (*aboral*) sur-

A large crown-of-thorns starfish under a ledge on the Great Barrier Reef.

face with hundreds of sharp, sturdy, venomous spines.

On one of my first dives off Australia, I found a crown-of-thorns that was about two feet in diameter. Fascinated by this animal I'd read about but never seen, I wanted to touch one of the spines — along its side, of course; this sea star was so huge that there was ample room for me to place a finger between the spines. As I reached toward the star, I heard shouting in the water: the Australian dive guide was trying to get my attention. When I turned to him, he vigorously signaled "NO!!" so I gave up my tactile quest.

Back on the boat I was warned by the dive guide and the captain that 1) the crown-of-thorns's body was covered with a mucus that was irritating to human skin, 2) I could have been envenomated if a spine had penetrated my skin, and 3) the crown-of-thorns was capable of **shooting** spines — one didn't have to impale oneself on the spines to be envenomated. All this was quite enough to make me want to read more and experience less of the crown-of-thorns.

Later I learned that the crown-of-thorns is **not** capable of shooting its spines out, but that it **does** have nasty venom and even its mucus irritates human skin.

On my next trip to Australia, on our way across the Barrier Reef, we dived an area that was badly infested with crowns-of-thorns. Several of our group were injured by them. The worst case — which fortunately wasn't as bad as it could have been — was a diver who got three spines imbedded near his knee, which bothered him for months afterward.

Like brittle stars, crowns-of-thorns avoid brightly-lit areas. In the Red Sea they were nocturnal. On the Barrier Reef either they weren't as bothered by light, or the competition for food somewhat outweighed their dislike of light, because individuals commonly fed in the daytime — but always in the shade, never in direct sunlight.

Crown-of-thorns sea stars use their tube feet for walking along the reef and for holding themselves onto vertical surfaces as they ingest coral polyps.

Almost every sea star I found in the Red Sea or the Coral Sea (including the Sea of Cortez) was accompanied by a pair of small shrimp whose coloration, like that of the shrimp in the basket star, perfectly matched that of the sea star. When I discussed the shrimp-in-star phenomenon with a biologist, he told me that it was very common for echinoderms to have crustaceans living with them. As I began to look at echinoderms more closely, I saw that the generalization seemed to be valid (with the exception of the sea cucumbers), particularly with the urchins.

Sea Urchins *(Class Echinoidea)*

Tiny elongated shrimp often hide along the spines of the long-spined urchin (*Diadema antillarum*), and small white crabs hitch rides through the sand on the undersides of heart urchins (*Meoma ventricosa*). My favorite crustacean-on-an-echinoderm relationship, perhaps because I've never

A Red Sea pencil urchin.

seen it described in any texts, is that of the little shrimp with the West Indi-an sea egg (*Tripneustes esculentus*).

The advantages to the crustaceans in these relationships are obvi-ous: they get free transportation, housing and protection, and perhaps even some food off the spines of their urchins. The advantages to the echinoderms are a bit less obvious, but may include having their spines cleaned. There don't seem to be any disadvantages. If one's guest causes no problems at all, why not be host?

Urchins have calcareous "skeletons" called *tests*; spines are attached to the test with ball-and-socket joints. Tube feet help urchins to move along and also help some of them to breathe.

A "regular" urchin, such as the longspined urchin or the West Indi-an sea egg, is radially symmetrical with the mouth on the underside and the anus on top. Between the mouth and the anus are five bands of holes for tube feet and spines. These urchins feed mainly on algae that they scrape off the substrate with five teeth, called the *Aristotle's lantern,* which open and close like drafting-type mechanical pencils close around their leads — and which can be completely protruded from the test, or com-pletely retracted inside it.

The other kind of urchins, the "irregular" urchins, are bilaterally symmetrical rather than radially symmetrical. This group includes the heart urchin and the sand dollar, whose membership in the phylum is evi-denced by the five-petal pattern on their top surfaces. They have under-slung mouths, so that as they move along, sand is plowed right into their

A short-spined Red Sea urchin, tube feet extended.

tests; they digest the edible matter and pass the rest. Heart urchins and sand dollars spend daylight hours beneath the sand. As night arrives they move onto and along the surface of the sand to feed.

Urchins have much less difficulty in getting food than they do in becoming food. Sea cucumbers are valued by some human cultures as food, but no marine animals seem to enjoy them. I haven't heard of anything that preys on sea stars or crinoids, and brittle stars are sought by only a few types of fish. Urchins, however, are eaten by people, by various kinds of fish (flounder and cod like sand dollars), by some mollusks (emperor helmets are able to get past the defenses of long-spined urchins), and, during the "urchin plague" of 1984, thousands of longspined urchins around Florida, the Caribbean, and the Bahamas were killed by a "mystery" disease.

Before dive guides got more knowledgeable about reef ecology, and well before the urchin plague, chopping up a long-spined urchin or two was a popular way for divers to attract fish — and was a good indicator of how valued urchins are by normally-shy queen angelfish, always-aggressive-but-not-quite-big-enough-to-open-urchins-by-themselves wrasses, triggerfish, filefish, and lots of others.

These fish, along with bristleworms (against which healthy urchins could easily defend themselves) feasted during the week or so the urchin plague took to decimate Bonaire's urchins, attacking sick *Diademas* as soon as their spines began to fall off.

Without their spines urchins could never survive, so it seems particularly unfair that their bad reputation is mainly caused by the damage their spines are capable of inflicting. The urchins are a case of misnamed echinoderms whose misnaming is more subtle than calling a sea star a "fish" ("starfish") or naming a sea cucumber after a plant. How can it be proven that the sea urchin's intent is not harmful, as might be the intent of a "land urchin"?

In Defense of Sea Urchins

First of all, the spines of most of the over seven hundred species of sea urchins aren't even sharp. These critters have common names like "slate pencil urchin" and "club urchin."

Second, the spines of the most maligned of all urchins, *Diadema*, have barbs (actually scales) that are aimed in the **wrong** direction; that is, they fight **against** the spine being imbedded, rather than against the spine being removed!!

Third, and probably most important: urchins **never** jump up and stab people. They are just there, and it is up to divers or other marine animals to impale themselves on the urchin's spines (assuming the spines are sharp enough).

I rest my case.

FISH

The most numerous marine vertebrates are the fish. There are well over 20,000 species of fish, which is more species than all other vertebrates combined. Since sixty percent of fish species — 12,000 — are marine, divers shouldn't feel bad if they can't identify every fish they see!

The fish are divided into two groups: cartilaginous fish (*Class Chondrichthyes*) and bony fish (*Class Osteichthyes*). The cartilaginous group includes the sharks and rays (subclass Elasmobranchii) and also a bizarre

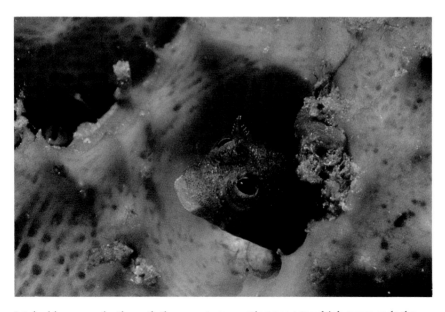

A tube blenny peaks through the orange encrusting sponge which surrounds the opening of its tube.

group of very deep-water fish called chimaeras.

Two major differences between cartilaginous fish and bony fish are in their scales and in their methods of buoyancy control. The scales of sharks and rays are actually tiny teeth, logically named dermal denticles. Like the overlapping plates that make up the scales of bony fish, dermal denticles face the posterior end of the shark; there's a noticeable difference in the feel of a shark's skin when it is stroked from head to tail, and from tail to head. The relative roughness or smoothness of an elasmobranch's skin depends on the species, though rays generally have smoother skin than sharks.

Bony fish stay in position in the water column by means of an internal buoyancy control device called the air bladder (also called the gas bladder or swim bladder). Experts in motionless hovering include trumpetfish, porcupinefish, trunkfish, and an occasional divemaster. Fish that don't use their gas bladders — some of which don't **have** functional gas bladders — include scorpionfish, morays, and lizardfish.

Elasmobranchs have no swim bladders, a fact that matters little to most rays since they rest on the bottom much of the time. Some sharks can adjust their buoyancy by gulping air into their stomachs; others are helped toward neutral buoyancy by large amounts of lighter-than-seawater oil in their livers. Because a shark's body is heavier than water, many sharks swim continuously to keep from sinking to the bottom.

Senses

Fish smell (and taste) the water through two canals on each side of their heads, above their mouths and in front of their eyes. Within these canals are taste buds, which send messages to the large olfactory lobes of their brains. Fish use this information for more than eating. Some scents (pheromones) are used to attract members of the opposite sex; other fish secrete scents that send messages of panic and/or danger.

The visual sense of fish is also well-developed, but fish see like the aptly-named "fisheye" lens. Fish cannot see the area just in front of their "noses," because their eyes are on the sides of their heads — but they can see separate side images with each eye.

Fish hear by sensing vibrations in the water. They do this in two ways: The first is through the membranous sacs that compose their ears. The second method of sensing vibrations is through sense organs humans do not have, their lateral lines, which are composed of pores leading to a canal lined with sensory cells. The lateral line enables fish to detect disturbances in the water, such as the flailing of injured animals or uncoordinated divers. It's a sense of "distant touch," and is different from their ears in that it can more accurately determine the direction of the disturbance. The lateral line sense is also believed to be the sense by which schooling fish keep their positions in relation to each other.

Reef whitetip sharks are able to breathe regardless of whether they're moving or not. Coral Sea.

Obtaining Oxygen

A fish's gills are extremely thin tissues containing many blood vessels, where oxygen is absorbed and carbon dioxide given off. Gills are efficient absorbers of oxygen; our lungs remove about 20% of the oxygen from the air we breathe, while a fish's gills remove 80% of the oxygen from the water that passes over them. On the other hand, air contains about 20% oxygen, while water holds only 2-3% oxygen, so the oxygen-absorbing efficiency of gills is a necessity rather than a luxury.

Since fish breathe so efficiently **in** water, it seems curious that they can't breathe **out** of water. The reason is that water pressure is necessary to hold the gill filaments apart — out of water the gills collapse against each other and aren't able to absorb oxygen.

The gills of a bony fish are on each side of the body behind and below the eyes. The fish moves along with its mouth open at least a bit, the gill flaps (*opercula*) flap, and water is drawn over the gills. Most fish can breathe even when they aren't moving by fanning their opercula.

The group of fish that is most famous for not being able to breathe unless they're moving are the sharks (who, as mentioned earlier, must swim or sink anyway because of their negative buoyancy). Elasmobranchs don't have opercula to protect and aerate their gills, but rather have five to seven gill slits on each side of their heads (or, in the case of the rays, on the ventral surface of their bodies). Only a few species of sharks, including

Southern stingray, dorsal view, showing eyes and spiracles, Bahamas.

nurse sharks, can breathe without swimming.

The Southern stingray (*Dasyatis americana*) is characteristic of rays in that it is dorso-ventrally (that is, from top to bottom) flattened, and spends its time foraging or resting on sandy areas. Its mouth faces the bot-

A green moray under a ledge, Bonaire.

tom — an inefficient water intake for breathing — so instead of inhaling water through its mouth, this ray has a special pair of water intakes (called *spiracles*) on the dorsal surface of its body. The water is exhaled through standard gill slits on the underside of the ray's body.

Flounders, like rays, are flattened fish that spend much time resting on the bottom. However, flounders are bony fish so they have typical bony fish gills protected by opercula. Flounders are laterally (from side to side) rather than dorso-ventrally flattened, and are physiologically lying on on their sides. The mouths of flounders face forward rather than down, so they are able to inhale water in the normal bony fish manner.

Because they constantly open and close their mouths just as other fish do, and because they show much more formidable teeth than other fish when they do so, moray eels were long thought to be threatening to divers. We now know that what they're doing is merely pumping water over their gills!

It's not often, luckily, that obtaining oxygen is confused with predatory behavior.

THE PREDATORS AND THE PREY

It's fascinating to look at a coral reef from the perspective of who eats what and how.

The clues are in the bodies of fish: form and function go together. So a sleek, hydrodynamically-efficiently-shaped fish must need to swim quickly, either to catch prey or avoid predation. Conversely, a lump-shaped fish must not have to worry about escaping predators or catching

THE FAR SIDE   By GARY LARSON

Burying itself deep in the mud, the hominideatodon, an evolutionary wonder, would slowly raise and lower its unique appendage in the hope of attracting its favorite prey.

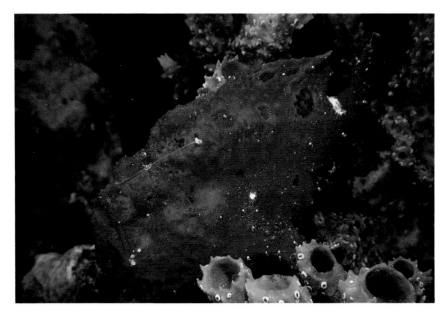

The longlure frogfish uses its pectoral and ventral fins to brace itself on the bottom or, in this case, a sponge; its unextended lure rests near its dorsal fin. The frogfish's mouth faces upward, which is the direction in which it lunges for prey.

fast-moving prey.

A fish with a small, pursed mouth would have to eat small prey; a large mouth indicates larger prey. If a fish's mouth faces downward, the fish is likely to be a bottom-feeder; if it faces upward, the fish may spend time **on** the bottom feeding on things above it.

All of these generalizations are perfectly logical; one needs only to think about them. Now let's use these kinds of clues to figure out the lifestyles of specific fish.

We'll start with the frogfish. **The clues:** It's lumpy-dumpy in shape; conclusion: it must not need to swim quickly. It has a relatively large mouth; conclusion: it eats largish prey. Its mouth faces up; conclusion: the frogfish eats prey that is above it.

The facts: The frogfish, an ambush predator, makes its living imitating sponges. Lest it be betrayed by its breathing, the frogfish's gills are covered with skin, and exhaled water comes out an opening in the equivalent of the fish's armpit. The frogfish is sponge-colored and rests on the bottom, or on an actual sponge, moving only its *illicium* (the first dorsal spine, modified into a fishing lure). When a prey fish, such as a wrasse or chromis, is attracted by the lure, the frogfish sucks it down into its expandable stomach. I've seen a frogfish swallow a fish that was almost as big as the frogfish itself!

Most frogfish will eat only live fish, but I was able to persuade one individual (whom I named Lazybones) to eat dead fish. The interesting

Diver Sandy Levy spots a frogfish on this piling of Bonaire's Old Pier — can you find it?

thing about these feedings was, despite the fact that my offerings were about the same size as Lazybones' normal prey, the frogfish couldn't swallow the entire baitfish; I'd have to snap off the tail of the baitfish to enable Lazybones to close his mouth! Apparently the struggling of a living fish actually enables the frogfish to swallow it.

Frogfish don't have to worry about escaping predation, since they look so much like sponges it's unlikely any predator could differentiate between them and a sponge — and since sponges are generally not eaten by marine animals, looking like sponges provides full protection to frogfish.

Next: Lizardfish. **The clues:** Like frogfish, lizardfish spend much of their time resting on the bottom, and their mouths face upward; conclusion: they eat prey that's above them. They have much sleeker bodies than frogfish and aren't so well camouflaged; conclusion: they need to swim quickly. They have much more prominent teeth than frogfish; conclusion: they use their teeth to hold struggling prey.

The facts: Lizardfish are chase-predators. They rest on the bottom until they see a fish moving into a position (not necessarily close to them) where they can grab it; they zoom up in a route they've predetermined, bite where they expect their prey to be, and return with their prey or empty-mouthed.

Porcupinefish. **The clues:** The porcupinefish is roundish in shape and moves slowly; conclusion: it doesn't have to avoid predators by outswimming them. Its teeth are fused together into plates; conclusion: it crushes its food.

This orange seahorse has a light coating of algae, like the rock it holds onto. ▶
Bonaire.

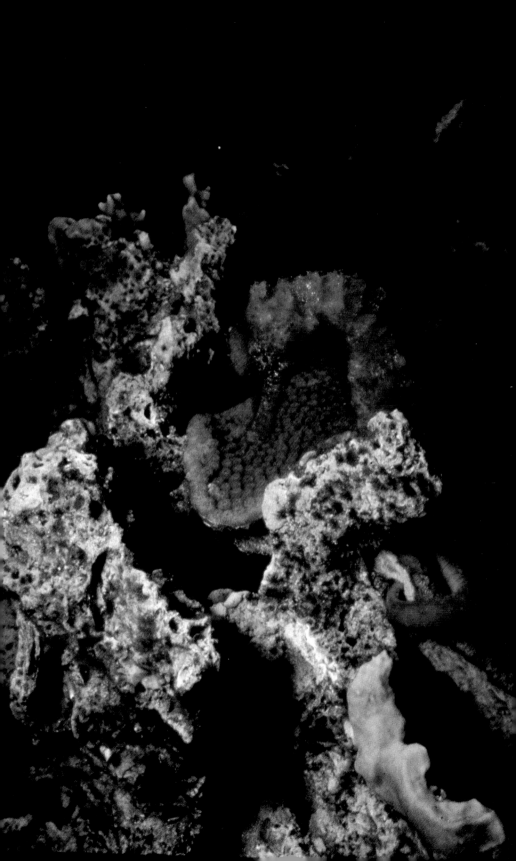

The facts: A porcupinefish can discourage unwelcome advances by puffing up and erecting its spines; it then becomes a very large, unappetizing, prickly thing, which most predators do their best to avoid. Its fused teeth enable it to crush the shells of crabs and mollusks.

Last but not least: the seahorse. **The clues:** Seahorses swim very slowly and inefficiently (even the slowest marine *Homo sapiens* can catch up to seahorses): conclusion: they don't have to swim quickly to avoid predation or to capture food. Seahorses have straw-like, pursed mouths; conclusion: they eat tiny creatures like plankton.

The facts: Seahorses have no specific predators (except for generally omnivorous fish and occasional underwater photographers). Seahorses nevertheless remain unobtrusive by imitating algae or sponges. As marine aquarists know, seahorses eat brine shrimp in captivity and similar planktonic critters in the sea.

Other fish that have interesting adaptations for feeding include longsnout butterflyfish, which use their long snouts to pluck invertebrates from deeper crevices than other fish can reach; goatfish, whose below-the-mouth barbels stir up the sand so their downward-facing mouths can gobble up any disturbed creatures; and parrotfish, whose fused teeth enable them to scrape up algae and coral polyps (you can hear parrotfish scraping under water).

One of the most unspecialized fish on Caribbean reefs is the yellowtail snapper, which is smart enough to learn what marine *Homo sapiens* "fish-feeding" posture looks like and quick enough to get all the food before most other fish even know the buffet line has been opened! This opportunism on the part of Bonaire's yellowtails, in conjunction with lots of divers feeding the fish, has resulted in a greater-than-normal population of larger-than-usual yellowtail snappers — an interesting diver-caused phenomenon.

Another fish likely to get the attention of divers is the damselfish, since these pugnacious little fish attack divers, parrotfish, barracudas, and anything else that comes near their territories and the algae gardens within them. Since a single fish trying to nibble at a damselfish's garden is sure to be chased away, surgeonfish have developed a better method of poaching: a school — actually, more like a mob — of surgeonfish descend together on a garden. One or two of the surgeonfish are attacked by the garden's owner, of course, but this enables the rest of the mob to feast on the garden undisturbed.

The situation is further complicated for the poor damselfish by trumpetfish, ambush-predators of small fish, who often join the mobs of surgeonfish for the purpose of catching preoccupied damselfish as they frantically protect their gardens.

When trumpetfish aren't taking advantage of the diversion created by surgeonfish mobs, they often use other fish, such as parrotfish, Spanish hogfish, or groupers, as stalking horses. The trumpetfish lines itself up

The only way I could get this close to a damselfish that wasn't attacking me was at night. Fiji, 1:3.

along the side of the top of the other fish and goes wherever the other fish goes. Prey fish may notice only the larger fish and not the trumpetfish.

If the "horse" itself is a predator, the trumpetfish can suck up small fish fleeing from that predator but not noticing the trumpetfish; if the "horse" is a grazer like a parrotfish, the small fish are vulnerable precisely because they're **not** frightened.

Moray eels are long and skinny like trumpetfish, and like trumpetfish they have some behaviors that can't readily be deduced from their appearance.

The long flexible bodies of morays tell us they can move around in areas of limited space. Their sharp teeth proclaim a need to hold struggling prey. Their long external nostrils show that they hunt more by scent than by sight. (These conclusions are valid for spotted-moray types in particular. Other morays, like the chain moray and the green moray, have shorter jaws and less sharp teeth than the spotted moray because they eat more crustaceans and mollusks than they do fish. Like the spotted moray, they hunt mainly by scent and use their flexibility to move around in tight spots.)

Morays have a couple of behaviors, though, that their appearance doesn't betray: their spinning behavior and their knot-tying behavior. The first time I saw the spinning I was wading along a gently sloping rock shore of an island in the Coral Sea. A moray eel grabbed one of the legs of a shore-zone crab and spun around on its long axis until it twisted the crab's leg off! It swallowed the leg, caught the crab again and twisted another leg off; then the crab escaped into a crevice. I've never seen a moray exhibit that behavior any other time, but snake eels — of the same basic body shape as

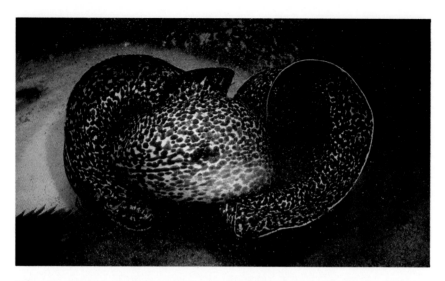

When a moray wants to eat something too big for a single swallow, one method it uses is to bite into the food with its long teeth, wrap its tail around its body, and pull its head through the resulting knot — thus gaining the leverage to tear away a bite-sized piece of the food.

morays — regularly spin to twist off bite-sized pieces of the food I offer them, or to twist the food into a better position for swallowing.

Moray eels are scavengers as well as predators, and will readily eat dead fish such as that discarded by fishermen. If a moray finds a piece of fish that's too big for it to eat in one bite, the moray flips its tail around its body, creating a knot, and pulls its head through the knot. The eel's long sharp teeth keep a tight grip on the food, and the coils of the eel's own body provide it with the leverage it needs to rip a piece of the food away! It swallows that piece, takes a mouth-grip on another, and repeats the process until the food is gone or it's so full it can't eat any more.

Avoiding Predation

The ingenuity of undersea predators is matched by the ingenuity of their prey.

"Safety in numbers" is the rule that schooling fish bet their lives on, for two reasons: one is a statistically lower chance of being eaten and the other is that a tightly-packed school of fish, especially in poor visibility, looks like one huge undersea animal — which few predators are crazy enough to challenge!

Other fish, including groupers, moray eels, and spotted drums, find havens in crevices and caves. Tiny blennies live in worm tubes. Clingfish cling to the undersides of rocks and ledges. Various species of gobies live safely in sponges.

Even an apparently barren sand bottom can provide shelter: garden

A school of grunts, Little Cayman.

eels have mucus-lined burrows in the sand from which they extend but never leave. Razorfish plunge into "dive sites" in the sand, and stay there until satisfied that the threat is gone. Snake eels use the sharp points of their tails to dig backwards under the sand until they're completely buried. Sand tilefish create dens out of intricately-piled rubble.

Yellowhead jawfish make vertical burrows in the sand, line them with rocks and retreat into them at night or when threatened. This seems like good protection, and it usually is, but I remember one incident when the jawfish's burrow was more of a danger than a shelter. My buddy and I realized something was amiss when I passed close above a yellowhead jawfish and it didn't sink out of sight. We had found an inhabited jawfish burrow; nearby was a second burrow, also inhabited — by a spotted moray! The jawfish had managed to escape as the little eel violated its burrow, but now the poor jawfish was homeless. I tried to encourage it into its companion's burrow, but the first jawfish protected its home even from its needy compatriot (or mate).

I was feeling less heartless, so I lured the moray out of the jawfish's home with a piece of fish; once the moray was completely out of the burrow, my buddy and I encouraged it to find another home. The jawfish returned to its burrow, neither it nor the moray any worse off as a result of their mutual misadventure.

Other potential prey creatures use camouflage to avoid their predators. The vertical bars of the banded butterflyfish help it to disappear within staghorn coral or gorgonians. Damselfish, small and neutrally-colored,

These Fijian butterflyfish have three stripes; the fact that one passes through their eyes is no coincidence.

easily blend in with their backgrounds. Sand-colored gobies scurry around invisibly on the sand. Peacock flounders, flattened against the bottom, match the colors of sand or algae-covered rock almost perfectly.

One marine environment with incredibly well-camouflaged animals is sargassum weed — not when it's attached to the bottom, but when it floats at the water's surface in clumps. I used to investigate it from the seawalls bordering Miami canals, and I still examine it whenever I find it. In those yucky-looking clumps of weed live sargassum fish (small anglerfish related to frogfish), sargassum shrimp, and sargassum crabs, all perfectly mottled to match the weed in which they spend their entire (adult) lives. In addition to the sargassum-critters, many other fish and invertebrates spend part of their lives within the sargassum community: pipefish and seahorses, young triggerfish and filefish, even the juveniles of some blue-water fish. By looking like sargassum, or hiding within it, these fish attain relative safety from predators.

Another type of camouflage is used expertly by butterflyfish: hiding their real eyes and presenting false eyes to their predators. A chase-predator who perceives the prey's eyes at the wrong end of its body will expect it to flee in the wrong direction — and the potential prey has a greater chance to escape. All five species of Caribbean butterflyfish have dark stripes camouflaging their eyes, and the foureye and spotfin butter-flyfish (as their names emphasize) have false eyespots near their tails.

Some reef fish seem especially vulnerable to predators, yet remain safe: trunkfish, for example, and the tiny Caribbean sharpnose puffer. Their secret? They taste terrible! Time after time I've seen scorpionfish suck down sharpnose puffers and spit them out. The little puffers usually don't even bother to inflate themselves, so sure are they of their distaste-fulness. The trunkfish's defense is different, in that it has a distasteful

A close-up of a Red Sea lionfish.

secretion used only when it is particularly upset — like when another fish tries to gobble it down, or when it's introduced into an aquarium.

Other puffers, burrfish, and porcupinefish use their puffing defensively; they discourage predators by making themselves larger. The flying gurnard, a bottom fish that doesn't fly at all, uses the same principle — but by illusion rather than fact: When threatened, the gurnard spreads its "wings" (expandable pectoral fins), discouraging predators by suddenly appearing to be much larger. Lionfish use the same system, spreading their winglike (and venomous) pectoral fins toward a threat.

Filefish and triggerfish simply extend their triggers and become difficult to swallow. As a child I kept some sargassum weed and its inhabitants. One morning I found a dead sargassum fish with a dead filefish in its mouth. Apparently the sargassum fish had tried to swallow the filefish, the filefish had extended its spine, and the sargassum fish could neither swallow it nor spit it out. They both drowned.

More recently I observed a similar incident with a happier ending. I was feeding a number of fish, including a filefish, when a scorpionfish tried to gobble up the filefish (backwards, in this case). The filefish extended its spine, and the two were at an impasse with the filefish's spine braced on the outside of the scorpionfish's mouth. Eventually the scorpionfish spit out the filefish. The filefish zoomed away, and perhaps realized that it wasn't as invulnerable to scorpionfish as it had thought!

The ultimate defense would seem to be that of scorpionfish, stonefish, and lionfish: venomous spines. The tricky part of this defense is

◄ *Jacks frequently accompany foraging Southern stingrays, hoping to gobble down what the stingray disturbs.*

educating predators about the venom. If the predator dies after it's eaten its (venomous) prey, the venom hasn't done its owner any good. If, however, a predator is made very sick, it would perhaps leave other such venomous prey alone. It's also possible that predators instinctively avoid venomous prey, or that the totally fearless attitude of venomous fish is itself confusing, and thus discouraging, to predators.

Despite the fact that we tend to think of the coral reef as a "fish-eat-fish" world, divers are able to observe surprisingly little predation. I've watched barracudas and jacks swim through huge schools of bait fish, never succeeding at eating one single fish. I've watched trumpetfish display their full repertoire of stalking techniques in vain, as their prey, keeping only a casual eye on the stalkers to stay out of position, blithely continued nibbling or grazing.

I've watched wrasses, grunts, damselfish, and even a French angelfish or two apparently concentrating totally on the hot dog I offered — until I moved the hot dog within lunging range of a scorpionfish. Then, as if by magic, all the wrasses, grunts, and damselfish dematerialized, leaving only the too-big-for-scorpionfish-dinner French angelfish. When I moved the hot dog back out of range, the smaller fish **re**materialized instantly, only to **de**materialize again as I moved the food closer to their predator.

When the mobbed-by-wrasses hot dog was lowered near a frogfish instead of a scorpionfish, the wrasses were no less conscious of the predator, but could afford to be more careless since the smaller frogfish would have to grab them headfirst (hydrodynamically efficiently) to succeed in swallowing them. Often a wrasse would wave its tail in the face of the frogfish, or present its body to it broadside, as if to torment its predator.

The tail-waving and broadside-presentation of potential prey to their predator is a behavior I've also seen with French angelfish, Spanish hogfish, and other medium-sized fish in relation to moray eels, especially if the moray had just arrived at that spot. It seems as if the fish are trying to make the eel uncomfortable enough to leave, and often they succeed.

Reef watchers who expect to see lots of predation taking place forget how seldom most predators must succeed. After a frogfish gobbles down a fish almost as large as itself, it doesn't need to hunt again for more than a week. A trumpetfish or scorpionfish may need to capture only one or two small fish a week — the rest of its time is spent stalking those small fish or resting. Fish somehow know just when a predator is stalking; their job is to not get into a swallowable position. The predator-prey interactions are a dance of moves and countermoves.

One day I inadvertently attracted one predator and enabled another to acquire prey. My buddy, Dennis Gast, had wanted to photograph me feeding an eel, so we visited a spotted moray named Weird Werner. I gave

This trumpetfish's mouth is distended because it has just gobbled down a small fish. Bonaire. Photo by Ed Cielakie

Werner a hot dog; this being one of his weird days, he spit it out. I retrieved the hot dog and held it out to some wrasses and grunts and a young French angelfish. Dennis began to take photos of the little fish feeding. Werner retreated into his den. Soon a trumpetfish began to stalk the wrasses. To "aim" itself better, the trumpetfish backed up a bit — a bit too close to Werner, unobtrusively watching the action from his doorway. In a flash Werner grabbed the trumpetfish.

That trumpetfish had thought it was the hunter only to discover — too late — that it was also the hunt**ed.** The entire event lasted no more than 60 seconds, and makes it clear why divers rarely, if ever, see instances of predation in the tiny percentage of the lives of marine animals that they are able to watch.

Another reason divers rarely see predation is that much active hunting takes place when people tend not to dive: during the low-light, changing-of-the-guard situations of dawn and dusk.

REPRODUCTION IN FISH

THE FAR SIDE   By GARY LARSON

**The committee to decide whether spawning
should be taught in school.**

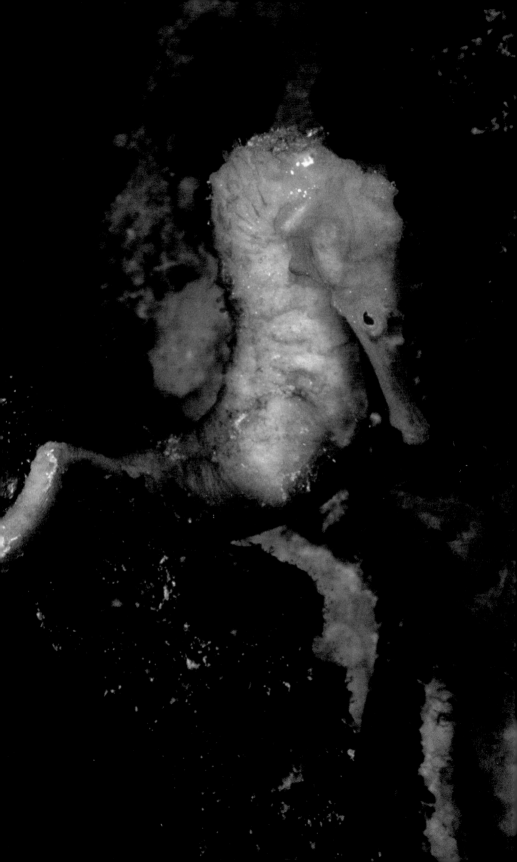

Most fish, like marine worms and other invertebrates, release thousands of eggs and sperm into the water simultaneously, trusting to proximity that some of the eggs will be fertilized and, of those, a few may survive to adulthood. Schooling fish generally spawn all together; the females release a scent along with their eggs that triggers the release of sperm by the males. The eggs float to the plankton layer where they enter the survival lottery.

Many reef fish spawn in pairs. The release of eggs and sperm is preceded by a courtship ritual that serves to synchronize the snap-release of gametes. The synchronization is what's important about the courtship from the point of view of survival of the fish; from the perspective of divers, the courtship is notification that spawning is about to occur.

That is, any time you see a pair of fish acting peculiarly, suspect that courtship is in progress.

The courtship of cowfish becomes noticeable when one fish's colors get so intense they're almost luminescent and the other fish becomes very pale. The two chase each other around for a while, then one (probably the male) places its mouth on the flat bottom of the other (probably the female), and pushes it upward in a spiraling path. This activity is visible from above the water as well as underneath, since cowfish actually break the surface as they complete their spawning rush.

Pearly razorfish also spawn in pairs, but unlike the cowfish, a male pearly razorfish is likely to mate with more than one female. Each male pearly has a territory that he guards from the males of surrounding territories and from other fish, such as small flounders, that might compete with the razorfish for food. Within the male's territory live the females of his harem.

One summer I assisted Dr. Eugenie Clark by observing pearly razorfish. I chose the largest harem I could find: one male and seven female pearlies.

In the late afternoon the bodies of the females would swell with developing eggs until a small red protrusion appeared at their genital opening. This red "balloon" showed — to observing divers as well as to male pearlies — that the female was ready to spawn.

As spawning time approached, the male circled throughout his territory from female to female, checking for red balloons and encouraging each female to join him in a spawning rush regardless of whether the red balloon was visible. Generally the females ignored these early attentions, but once in a while one would participate in an unconsummated spawning rush. Finally, each female would respond to the male and spin up to the surface with her body parallel to his. At the peak of the upward rush, they'd snap-release sperm and eggs, and then they'd return to the bottom — she, unswollen and red balloon-less, to resume before-bedtime foraging, and he to approach the next female.

As time passed, the females who hadn't spawned would become

increasingly aggressive toward the male, chasing him around and display-ing their red balloons to him. They seemed to get uncomfortable as their bodies swelled with eggs, and needed to spawn to end the discomfort.

When I describe pearly razorfish harem activities, some joker in the crowd usually snickers and sighs about the lucky male who got to spawn seven times every day — but his envy evaporates when he learns that the male **had** to service his entire harem regardless of whether he was in the mood. If he couldn't manage his responsibilities, he'd lose his job: a neigh-boring male would steal one or more of his females, or his dominant female would change to a male and split the harem! (This in fact seemed to have happened, for a few months later that large harem was gone.)

Razorfish first reach sexual maturity as females, and only under specific circumstances — such as the male succumbing to a predator, or not being able to service all of his females — can one of the females actu-ally change sex and become a male.

As strange as that sort of sexual biology may seem to us stodgy humans, the reproductive sagas of other wrasses are even more bizarre. Some wrasses reach sexual maturity relatively "normally," as either males or females, with the initial-stage males looking almost exactly like the females. Generally, these initial-stage males spawn with females in groups, although since the males and females look so much alike, the groups occasionally consist only of males!

In addition to the initial-stage males, there are also large, differently-colored males who started life as females and changed to males. These ter-minal males (occasionally called supermales) usually spawn with females on a one-to-one basis. Sometimes initial-stage males join, uninvited, the spawning rushes of terminal-stage males, sneakily adding their sperm to that of the terminal male in the contest for the female's eggs. Sometimes the terminal males end up spawning with initial-stage males instead of females, since the initial-stage males look so much like the females! And sometimes the testes of the terminal-stage males produce non-viable sperm anyway, so that whatever spawning efforts they make can't result in offspring. Perhaps the wrasses understand what's going on, but to humans many aspects of their sex lives are still unsolved puzzles...

The reproductive puzzles of the hamlets seem to be solved, though: they are simultaneous hermaphrodites — that is, each fish is a fully func-tioning male **and** female at the same time! Hamlets spawn at twilight, so the few times I'd observed them had always been accompanied by eye-strain — which was, at first, what I thought was the cause of my confu-sion. When I began watching a pair, one hamlet had a bold bar along its side and the other was pale in coloration. They "danced" around a bit, joined in a spawning clasp, released eggs and sperm and each other, swam slowly back down to the coral head over which they were spawning, and repeated the procedure. The confusing part was that I could have sworn that the second time they switched color patterns — the one with the bold bar had turned pale and the pale one had acquired a bar, and their positions in the clasp were correspondingly reversed.

It wasn't until later, when I read through some of Dr. Eugenie Clark's work at Mote Marine Lab and Dr. Ronald Thresher's book *Reef Fishes,* that my confusion was resolved: each hamlet has both male and female reproductive organs. In a given pair, the one that is the most swollen with eggs functions as the female first, but each acts as both female and male to complement the other.

One of the longest courtship and spawning interactions I've seen took place between a pair of emperor angelfish at twilight, in the Coral Sea. When we saw one of these usually-solitary fish join another, my buddy and I sacrificed the photo opportunities to see what would happen.

Unfortunately for the angelfish, other critters interrupted their interactions that evening: no sooner had they begun their courtship than a huge jack swam menacingly through the area. The angelfish darted back under a crevice. For the next twenty minutes or so we watched the angelfish begin their courtship again and again, only to be interrupted again and again by the jack or by reef sharks. Finally, just before dark, the coast cleared and the two ardent angels rushed toward the surface, snapped out their eggs and sperm, and dashed back to the safety of the bottom.

Angelfish are among the majority of reef fish: they have nothing to do with their eggs after spawning. A few other fish, including the damselfish, provide some parental care. One of the reasons damsels are so aggressive is that they're protecting their algae gardens, but another reason is that they may be protecting their eggs.

Take the sergeant-major, for example, a damselfish that lives in virtually all warm seas including the Caribbean, the Indo-Pacific, and the Sea of Cortez. When spawning time approaches, the male sergeant-major finds a flat site to his liking — such as the urchin-eaten underside of a coral head, or the side of a wreck — and dusts it off. He does a looping dance to attract a female, who swims in tight head-to-tail circles with him as she lays a single layer of purplish eggs, each about the size of a pinhead, on the nesting site. The male fertilizes the eggs, and if the female hasn't left by then he chases her away. The male will attract additional females to his nesting site if he can, and repeat the procedures.

The eggs change color as they develop from purplish to greenish to grey, and the male sergeant-major guards them from predators such as wrasses and even mollusks like the murex. One of the sneakiest sergeant-major egg predators I've seen was a tiny crab who lived in a tiny hole in the middle of a nest. The area surrounding the crab's burrow was denuded of eggs, because periodically a little claw would reach out and casually snitch an egg!

Usually, though, male sergeant-majors protect their eggs more efficiently than that. I often visit a spot on which two sergeant-majors nest, and place my underwater slate (which is white, so they can't miss it, and

A garibaldi, related to the sergeant-major, guards its eggs off Ana Capa Island.

light, so it doesn't hurt the eggs) near the eggs. One father, the one who behaves most typically, pushes at the slate with his mouth or waves his tail at it until it finally drifts off. The other father, more adaptable, soon figured out that he could grab the slate by its lanyard and carry it off his eggs! (He got a field promotion to lieutenant.)

California's bright orange garibaldi guards its eggs even more tenaciously than its sergeant-major relatives. I tried to place a small shell on a garibaldi nest one time, but before I had even let go of the shell the garibaldi attacked my hand so ferociously I suddenly remembered an appointment elsewhere!

Although the eggs of damselfish are better protected than those of "broadcast" breeders like cowfish and wrasses, other fish take even better care of their eggs.

Yellowhead jawfish look to me like little princesses, with blonde hair, lovely silver-blue gowns, and great big black eyes. The female lays her eggs in a burrow, and after the male fertilizes them he gathers them up into his mouth where he keeps them almost continuously until they hatch. An egg-protecting jawfish is fairly obvious: his cheeks are puffed out with eggs so large (bigger than their guard's eyes) that he can't completely close his mouth!

The undersea critter whose breeding habits are best known is the seahorse, and, as usual, the male carries the responsibility (and the eggs). The female seahorse lays her eggs into a brood pouch on the male, where they stay until they hatch. Then, with great muscular contortion, he ejects the tiny seahorses into the world.

Sharks and rays, cartilaginous fish, mate quite differently from the bony fish. Sperm is transferred to the females internally by means of specialized structures, called *claspers,* on the pelvic fins of the males. Because of this internal fertilization and the excellent protection provided for the eggs, sharks and rays can produce relatively few eggs and still survive as species. The eggs of some elasmobranchs are protected by horny cases (skate egg cases, common in temperate waters, are called ''mermaid's purses''), which entangle with algae and other bottom growth; eventually small, fully-formed young emerge.

Other sharks, like the nurse shark, are ovoviviparous; that is, the eggs develop and hatch in the female's reproductive tract. In some ovoviviparous sharks the internally-hatched embryos feed on fertilized eggs or other embryos until they are born!

Hammerheads and tigers are among the viviparous sharks: the young grow inside the female's uterus, receiving nourishment from both the egg yolk and the mother shark. This system is similar to the true placental situation in mammals, but it is not the same; there is no placenta in sharks. Instead, shark embryos start with huge amounts of yolk.

Much still isn't known about reproduction of marine fish. For example, only one egg case has been found for the whale shark (as of the time this was written). Observations have been made of writhing groups of snake eels at the water's surface on some nights, and conclusions drawn that this was spawning behavior — yet I've seen and heard of others seeing writhing masses of snake eels on the **bottom,** and concluded that **this** was spawning behavior! In general, if a fish is not particularly useful — or dangerous — to people, there isn't financing available for its study.

Which means that divers who want to learn more about marine animals have an open field — uh, ocean — waiting to teach us.

RELATIONSHIPS

Cleaners

Most marine animals relate to each other primarily when they're trying to eat, to avoid being eaten, or to reproduce. They have a few other types of relationships, though, and one is that of the cleaner stations.

The cleaner station is a uniquely visible example of underwater mutual aid. The cleaners get a meal of their clients' parasites: The clients get rid of their parasites. The only losers are the parasites, and no one cares about them anyway!

The faith that fish have in cleaners was illustrated for me in the Red Sea, when my attention was attracted by a grouper vigorously shaking its head. When I sneaked up on it to get a closer look, I saw the tail of a moray protruding from the grouper's gill flap! Apparently this grouper had seen the small moray out in the open and tried to eat it — but, at the last moment, the eel whipped its tail around and hooked it through the grouper's gill. The eel wasn't able to back out, and the grouper wasn't able to finish swallowing the eel. Impasse.

When the head-shaking proved unsuccessful, the grouper tried a cleaner station. The station's attendant wrasses couldn't help their client out of its plight, though; they didn't even notice the eel's tail sticking out of the grouper's gill.

Finally the grouper left the cleaner station, and, with an exceptionally strong twist of its head, broke the eel's hold and was able to finish swallowing it.

The grouper's instinctive response to discomfort was to visit a cleaner station.

It's fun to watch fish at a cleaner station. They hover in apparent enjoyment as the cleaners roam all over their bodies, including into their mouths and out their gill flaps. There's a non-aggression pact between cleaners and predators. I've never seen a predator eat a cleaner, and, obviously, the cleaner/client relationship wouldn't last long if predators ate their cleaners!

Fish being cleaned will often allow divers to get much closer than usual. This may be because the clients enjoy the cleaning so much they're willing to take a bit of a risk to prolong it, or it may be because they know that if real danger threatens, the cleaners will warn them by leaving. Whatever the reason, the cleaner stations provide wonderful opportunities for fish portraits, whether with "nature's camera" (the eye) or with an actual camera.

On one occasion a grouper spread his gill flaps so wide that I could see through his mouth and back out of his gill flaps to the reef behind him! On another occasion my buddy and I watched gobies swim into the mouth of a trumpetfish, and then watched them work **inside** the trumpetfish through the translucent skin on the fish's head!

Quite a bit of research has been done on cleaner stations.One scientist, examining the stomach contents of a number of cleaners, found relatively few parasites. The scientist theorized that fish patronize cleaner stations even when they're not bothered by parasites because the cleaning experience is pleasant. When the fish do have a problem with parasites, though, it's rapidly taken care of because their cleaner station appointments are established.

What benefit do cleaners get from a client with no parasites? Apparently they're satisfied with an occasional piece of dead skin or a stolen scale.

There are two types of cleaners: fish and crustaceans, usually shrimp. Most full-time fish cleaners, such as the Indo-Pacific's cleaner wrasses or the gobies of the Caribbean, display a "guild sign" of horizontal stripes. Some juvenile fish, including the juvenile French angelfish and Spanish hogfish of the Caribbean, also clean, but they don't display the

This Red Sea grouper had gobbled down a moray — but the eel hooked its tail around the grouper's gill flap. The grouper went to a wrasse cleaner station for help.

guild signs of full-time cleaners. The juvenile French angelfish advertises its cleaner status by swimming in a diagonal posture with exaggerated movements; when the angelfish outgrows its cleaner stage it swims in a straight-up position with simple side-to-side motions of its tail fin.

Shrimp cleaners generally display the guild sign of long white antennae, like those on the scarlet lady cleaner shrimp. Shrimp cleaners advertise by swaying their bodies from side to side and sometimes by wriggling their antennae.

Different marine locations have a different predominance of cleaners. I've seen many cleaner wrasses in the waters of the Coral Sea and the Red Sea, but no cleaner gobies and no cleaner shrimp. On Bonaire we seem to have more cleaner shrimp than cleaner fish, but the fish are mostly gobies and juvenile French angelfish, with a few wrasse stations. In the waters off Little Cayman I was amazed at the numbers of goby cleaner stations; every barrel sponge seemed to have a big population of cleaner gobies.

Unless they're as obtrusive as Little Cayman's gobies, the easiest way to find a cleaner station is to spot a fish being cleaned, and the easiest way to spot a fish being cleaned is to find one whose actions aren't as usual. Oops, isn't this the same as the hint for finding courting or spawning fish? To differentiate spawning fish from fish being cleaned, note that spawning fish usually move around a lot, and all the fish involved (usually two) are of the same species; fish being cleaned are usually **not** moving around much, and the cleaner fish are of a different species, and quite a bit smaller, than their clients.

Here are some behaviors of specific fish at cleaner stations: creole wrasses, those bluish-purple fish that hang out in the water column around the drop-off, are frequently seen in a head-down, mouth-open-and-extended position, with small yellow wrasses attending them. Parrotfish, when being cleaned by fish, take a head-up position and use their pectoral fins to stay in place; when they're being cleaned by shrimp they rest on the bottom. Groupers prefer cleaner stations under ledges, and often roll from side to side in apparent ecstasy. They also change color from pale to dark or vice versa, perhaps to make their parasites, camouflaged to match their normal coloration, more obvious to the cleaners (the groupers can change colors faster than the parasites can).

Other fish that change color at cleaner stations include goatfish and yellowtail snappers, both normally white with a yellow stripe, which both turn pinkish when being cleaned.

Goatfish and some groupers, judging by the frequency with which they patronize cleaner stations, are either the most clean or the least clean fish on the reef!

Moray eels are also cleaned frequently, but the cleaners seek them out rather than the other way around. Watch a moray that's swimming from place to place, and you'll see every neon goby it passes landing on it and scurrying around! Cleaner shrimp, even the banded coral shrimp, respond

Peter Gray has his hand cleaned by neon gobies. Little Cayman.

to morays equally enthusiastically. There must be something about the skin of a moray that attracts cleaners, despite the fact that the morays give the cleaners no encouragement and twitch irritably at their attentions.

Ghost cleaner shrimp, which are easy enough to see sitting outside their anemones waving their antennae back and forth, earn their name when they're cleaning: all that's visible is the client, often a hawkfish or goatfish, resting on the bottom. Only exceptionally careful observation reveals the small, almost clear shrimp moving around over the fish.

The most interesting way to watch cleaners at work is to be the client yourself, although getting cleaned can sometimes be a real challenge. I've had to adjust the position of my hand (to vertical rather than horizontal, since that's the normal orientation of most fish); I've had to remove rings; I've had to face into a current and even sway with a surge in some cases before I could get cleaners to react to my hand (the most fish-like part of my anatomy).

The cleaners I've found most willing to accept my hand are scarlet lady cleaner shrimp and Little Cayman's gobies. The only scarlet lady that's ever refused me lived under a rock with a spotted moray. I worked very hard to get the moray to face in the other direction so I could safely present my hand to the scarlet lady — which after all that trouble refused to clean me! I decided that the scarlet lady considered itself the personal valet of that moray, and wasn't interested in moonlighting.

The difference in the willingness of various gobies to clean is interesting. On Little Cayman when we presented our hands at the gobies' sponges, several gobies would immediately swarm aboard. The cleaner gobies on San Salvador were less forward, but still willing to clean — although occasionally I'd discover that while I was vainly extending my hand to one group of gobies, another group was working on my legs! On Bonaire the gobies only rarely accept a diver as a client.

Cleaner animals are divided into primary and secondary cleaners. The primary cleaners are defined as making most of their living from cleaning; the secondary cleaners clean as a second job. Primary cleaners are the most willing to clean, display the cleaner guild signs most obviously, and are least likely to be accidentally or deliberately eaten by a client. Divers are much more likely to see fish being cleaned by, or to be themselves cleaned by, primary cleaners.

Divers aren't the only marine animals to take advantage of primary cleaners: some wrasses and blennies are not actually cleaners, but rather **mimic** both the appearance and the actions of cleaner fish — until it becomes time to work. Then, rather than actually cleaning parasites and dead skin from their prospective clients, the mimic cleaners take the easier route of zooming over to the fish, grabbing a mouthful of perfectly healthy flesh, and zipping off to safety! I've observed fish at the moment of victimization: they twitch at the traitorous behavior of the cleaner and quickly depart the scene of their duping. Luckily, mimic cleaners don't readily accept "mimic clients," the hands of divers.

Tenants

One of the most interesting relationships between marine animals is between anemones and various species of damselfish, called "anemonefish," in some tropical waters (but not the Caribbean). The anemonefish survive not just in the vicinity of, but actually within the tentacles of their anemone landlords.

I had read about this relationship but never understood its fascination until I saw my first anemonefish, on the *Yongala* wreck in Australia. Within a large anemone, similar-looking to the common Caribbean anemone, was a pair of orange-and-white fish wriggling sensuously through the tentacles! The surprising part wasn't so much that the anemonefish could touch its

A pair of two-barred anemonefish, Fiji.

landlord's tentacles without being stung; to me what was amazing was the way these little fish seemed to actually bathe in the waving tentacles.

Studies of these relationships indicate that the bathing is necessary for the anemonefish to remain immune to the anemone's nematocysts. A fish that is removed from its anemone must reacclimatize to its host by gradually exposing itself to the anemone's stings. Apparently the mucus coating on the anemonefish becomes similar enough to the anemone's own mucus coating (or the anemonefish becomes covered with the actual mucus from the anemone's tentacles), so that the anemone no longer stings the fish, just as it doesn't sting itself.

The adult pair of anemonefish cleans the substrate around the anemone and lays their eggs under the protective umbrella of anemone tentacles. The fish guard the anemone, and the anemone "guards" their eggs.

Occasionally immature anemonefish live in the same anemone as a pair of adults, but the presence of the adults prevents the juveniles from maturing; they must either wait for one of the adults to disappear or find an uninhabited anemone, to mature.

During my first trip to Australia, I read of an experience of Valerie Taylor: when she fed an anemonefish, it fed the food to its anemone! This was something I wanted to see firsthand, so I began to offer food to various species of anemonefish — and found that the results depended on the species of fish. In some cases, the anemonefish eagerly grabbed the food from the water (or directly from my hand) and buried it within the anemone's tentacles. After five minutes or so of feeding every piece of food to the anemone, these fish would finally begin their own meals — proving that they weren't feeding the food to the anemone only because they wouldn't eat it themselves!

Other species of anemonefish would eat the food I offered but never pay any rent, and others neither ate nor paid.

The anemone-anemonefish relationship is clearly advantageous to the fish: They and their eggs, receive protection. The anemones tenanted by anemonefish who "pay rent" also benefit from the relationship. There are other benefits to the anemones, too: since the fish's eggs are laid

David Batalsky examines Anemone City.

beneath the anemone's tentacles, and the anemonefish (like the sergeant-majors) guard their eggs, they also guard the anemone from the occasional butterflyfish or other anemone-nibbler.

It's interesting that although this confirmed landlord-tenant relationship doesn't exist in the Caribbean, some young fish are definitely investigating anemones as potential landlords. In the last couple of years I've seen damselfish, wrasses, cardinalfish, a blenny or two, and even a redspotted hawkfish **inside** giant Caribbean anemones! In most cases the fish weren't making contact with the tentacles; when I fed the anemone the fish usually left it, and they don't bathe within the tentacles as true anemonefish do. Still, a few of these fish **are** making contact with the tentacles, and more than once I've seen damselfish sleeping in the protective ring of anemone tentacles.

There's no doubt that these youngsters are investigating the so-far-unused-by-Caribbean-fish ecological niche of anemones, and I suspect that, in a few thousand years or so, we'll see Caribbean anemones tenanted by and living with Caribbean fish.

THE TWENTY-FOUR HOUR REEF

With a good base of information about marine animals, all their behaviors can be put together into a multidimensional entity: the twenty-four hour reef.

Dawn

As light on the reef increases, the disappearing acts of the nocturnal invertebrates — urchins, crabs, mollusks — are completed. The reef appears motionless and empty.

One watching the right place on the sand might see the head of a razorfish as it emerges. If the watcher moves, the razorfish dives back under the sand in a flash; otherwise, groggy, it completes its entry into the water column and begins foraging. More razorfish appear.

A trip to the drop-off now reveals the usual diurnal schools of chromis, creole wrasse, and creole fish in the water column. Trumpetfish have left their vertical sleeping positions and begun to swim horizontally around the reef. Yellowtail snappers patrol the reef slope.

One minute the reef appears still, the next it is crowded with life. Day has begun.

Day

Daytime is when marine *Homo sapiens* are most likely to appear on the reef.

Groupers, yellowtail snappers, wrasses, and sergeant-majors crowd around divers in hopes of a free meal.

Parrotfish graze on coral and algae, audibly crunching away at coral heads and ultimately turning them into sand. Research in Bermuda showed that for every acre of reef, parrotfish created about a ton of sand each year!

Butterflyfish and angelfish, delicately nibbling, flit from coral head to rock to sponge.

Scorpionfish, lizardfish, flounders, frogfish, and other camouflaged ambush-predators wait for unwary prey to swim close enough to become a meal.

Brilliantly colored wrasses row from place to place with their pectoral fins, investigating reef crevices, the water column, and sand bottoms for food. They join goatfish and stingrays who stir up the sand in search of their own food, darting down to capture small animals the others miss.

The long white antennae of cleaner shrimp tremble to advertise their availability; juvenile angelfish and other juvenile cleaners swim with exaggerated motions so prospective clients know they're at work; primary cleaners, the gobies and wrasses need only appear at their stations to broadcast the message that the cleaner stations are open for business.

As the daylight hours pass, the bodies of some female fish swell with developing eggs.

Dusk

When the light begins to wane, activity on and around the reef increases briefly. Diurnal fish complete their foraging for the day. Those who will spawn begin their courtships.

Many predators, including sharks, barracudas, and groupers, concentrate their hunting energies during twilight hours; the frenzied activity of other fish may make their success more likely.

Some species of octopus also hunt most actively during this time — perhaps because there is enough light to help them find prey, but not so much light that their predators can see them. During razorfish-watching times, we occasionally saw, just before it got too dark to see at all, a small pale-colored octopus emerge from the sand! So far, the identity of those little octopuses remains a mystery.

As the light decreases, diurnal animals settle for the night. Razorfish and other small wrasses dart beneath the sand. Creole fish, creole wrasses, Spanish hogfish, and damselfish seek out crevices in the reef or the tubes of sponges in which to spend the night. Parrotfish, angelfish, and groupers settle on the bottom. The activities of trunkfish and cowfish gradually slow down until full dark finds them still hovering above the reef, but otherwise inactive.

Nocturnal fish become active. Spotted drums, blackbar soldierfish, cardinalfish, and others emerge from their heavily-shaded daytime shelters.

The great majority of invertebrates are active at night. It is then that the polyps of stony corals expand. Crabs and other crustaceans leave their dens to forage. Gastropod mollusks emerge from the shelter of the sand or from reef crevices. Long-spined urchins move out of their coral crevices to forage along algae-covered rocks; heart urchins and sand dollars move from beneath the sand to the top of it; the bodies of sea cucumbers expand; brittle stars creep out of the sponge cavities and deep ledges in which they spend daylight hours.

As darkness falls, the appearance of the reef changes.

Night

For those of us who dive because we want to understand the marine environment, night diving is essential.

I met a diver once who had marked off a beach site (less than 30 feet deep), dived it a couple of times in the daytime, got a stack of tanks, and proceeded to try to dive it all night. He didn't succeed, but he did spend seven hours or so in the water that night — alternating dives with warm showers.

His method is a pretty extreme way to get to know a reef on a twenty-four hour basis, but I admire his idea and his spunk (I have a few reservations about other things). My own so-far-unrealized goal is to pick out a reef area and dive it on a schedule: if I could dive it regularly every four to five hours, I could dive the reef all twenty-four hours in less than a week.

Luckily, such concentration on detail is not necessary to learn the general aspects of a reef (day or night) although it might end the debate among night divers as to whether diving very late at night is "better" than diving shortly after the reef gets dark.

My own answer to the question relates to the general behavior of reef animals at night: some things occur nightly and can be seen on just about any dive by everyone who knows how to look for them; seeing other things is more a matter of luck. During my experiences working on a night

This parrotfish's cocoon is obvious. Bonaire.

reef film, the only fairly consistent difference I noticed between the early and late night dives was that it seemed we saw more parrotfish in cocoons on the later dives.

Those animals who begin to appear at dusk are the "guarantees" of the nocturnal reef, the steady performers we can count on. As we slowly move along, studying the smallest creatures of the night reef, we occasionally have the pleasure of finding one that is not a guarantee but a pleasant surprise.

Most of the surprises I find are nudibranchs. Even after more than 3000 dives on Bonaire's reefs, I still find nudis that I've never seen before (and never seem to see again). One night I found a tiny black and cream colored nudi that settled comfortably on my ring, smaller than the width of the ring! Another night, in a coral crevice, I found a two-inch or so nudi the color and texture of a peeled tomato. That creature was a perfect size and color for macro-photography, but of course no one had a camera.

Bonaire's most common nocturnal nudibranch is identified by Zeiller *(Tropical Marine Invertebrates of South Florida and the Bahama Islands)* as the leathery nudibranch *(Platydoris angustipes)*, a dorid nudibranch whose common name describes its texture but not its beauty. This animal is soft red with white markings, and when I spotted my first one I thought it was the extended body of a tulip shell. Closer examination revealed two overlapping nudibranchs. Each nudi had a gill tuft extending from the top of its body, which retracted when I disturbed the nudi. I picked up one of the animals and was surprised to discover how stiff its body was — it felt like a piece of old leather — but after I replaced it it gradually flattened out again and the gill tuft reappeared.

Leathery nudibranchs spend their days hidden beneath coral or bottom debris, but even if they were fully exposed in the day they'd be more difficult to find because the blue-filtering action of the water would change their reddish coloration to grey or black. One of the fascinations of night diving is the vivid hot colors of the reef, many of which are there in the daytime but are not visible until a diver illuminates them with the white light of a dive light.

The white light of a dive light can also expose camouflaged animals such as frogfish and the shrimp in crinoids. The coloration of frogfish usually matches — but not perfectly — the sponges on which they rest. The difference in colors is not noticeable under filtered daylight, but in the beam of dive lights the frogfish is much more obvious.

Generally I spend my night dives moving slowly, very close to the coral or the bottom, examining the surfaces below me for small creatures. Periodically I shine my light in a wide circle in the hopes of discovering some larger animal at work. It was during these beam sweeps that I first noticed a fascinating characteristic of the eyes of some nocturnal fish, particularly blackbar soldierfish and tarpon: the wide beam from my light strikes their eyes and seems to be reflected back out as a narrower, brighter beam! Since both blackbar soldierfish and tarpon feed at night on small creatures, it would be an obvious advantage to them if their eyes were capable of focusing scattered ambient light into a beam they could direct.

Tarpon are wonderful animals to watch, but shining a light on them only frightens them away. The best way to watch these large, graceful animals is by ambient light, and the most fun I had doing this was in Belize. The boat crew insisted on shining a bright light off the stern of the boat when we did our night dives, and though the light disturbed much of the life on the reef, it attracted a lot of plankton, which attracted numbers of small fish, which attracted numbers of tarpon. At the end of our night dives, we'd hold onto the stern down-lines and drift back and forth with the boat, lights off, watching the tarpon. Since we were fairly motionless and not using lights, the tarpon seemed to consider us part of the boat; they'd swim within three feet of us, and sometimes I could feel the water currents they created as they passed. Fantastic!

If we turned around, facing away from the lighted water, and waved our hands, we could see the "glitter" created by the bioluminescent plankton we disturbed.

Only recently I discovered a trick performed by brittlestars that live beneath sand bottoms. At night these brittlestars extend their five sand-colored arms above the surface of the sand to feed. When I find a brittle-star, turn my light away from it and tap its arms — and the arms light up! Not only do they light up, but it looks as if neon light is moving through them. The illusion is that the arms themselves are retreating back beneath the sand, which they usually are not doing. These brittlestars have a handy defense mechanism against animals that eat star arms, and provide a harmless game for divers to play.

Living on the sand, in the same areas as these brittlestars, are the tube anemones (of the order Ceriantharia), with their extended columns and two circles of tentacles.

Also along the sand, especially on the interface between the sand and the reef, cornetfish *(Fistularia tabacaria)* can occasionally be found. Cornetfish are long and skinny like trumpetfish, but there are major differences between the two fish: cornetfish have a blue-spotted color phase, and trumpetfish never have spots; cornetfish have a whiplike appendage extending from the fork of their tails, which trumpetfish don't have; cornetfish never hover vertically; cornetfish swim with whole-body undulations, and trumpetfish swim using their dorsal and ventral fins. Also, cornetfish will occasionally grow to five or six feet long, much larger than trumpetfish.

Although cornetfish can be seen in the daytime, they can be approached much more closely at night. I've found that if a bright light is aimed directly at the eyes of a cornetfish and held steady, the fish will often become mesmerized and move slowly toward the light as if being drawn by invisible lines. One time a dazzled cornetfish swam between my neck and my hair!

Aiming a bright light steadily has also drawn barracudas and cow-fish toward me; when the barracuda got within about two feet it snapped out of the trance. Sleeping cowfish often allow themselves to be stroked, but be aware that those horns are a potent weapon. I discovered this when I was in the path of a cowfish that suddenly took fright, floored its acceler-ator, and crashed into my leg. Its horns easily penetrated my 3/16" wet suit, and left me with two puncture wounds that were soon surrounded by bruises — probably what I deserve for disturbing a soundly sleeping fish!

After my first few night dives I developed the self-control to leave most sleeping fish alone — my attempts to touch them usually failed and only resulted in the panicked fish swimming into coral heads or, worse, urchins. It became more interesting to try to figure out which fish I was seeing: at night butterflyfish acquire dark blotches, surgeonfish acquire light blotches, and blue tangs become striped!

I also became challenged by trying to find all the places in which fish slept. Butterflyfish and tangs and groupers and angelfish and parrot-fish were easy, since they rest right out in the open. More difficult were creole fish and damselfish, until I found a few cuddled into coral crevices and even in the tubes of sponges — which is probably why night divers occasionally see moray eels in the tubes of sponges. The most concealed nighttime sleepers are the smaller wrasses, who burrow down into the sand for the night. That habit causes great concern among aquarium own-

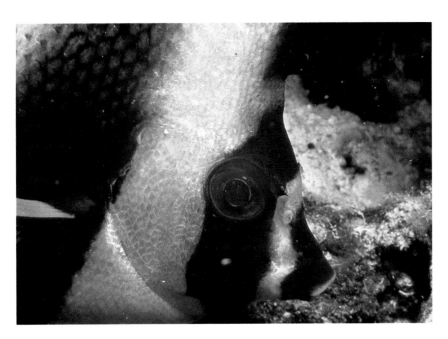

A sleeping butterflyfish, Fiji.

ers of recently-purchased (and nocturnally disappearing) wrasses, and it caused me great concern when one popped up from beneath my hand one night!

One of my favorite nighttime sights is a large yellow tube sponge community that I call "the critter condominium." One night I found there, cuddled among anemones and brittlestars and other sleeping fish, a small filefish. I noticed that when my buddy created a current, the filefish's body swayed but its mouth stayed in place — it was biting a projection of the sponge as it slept! Since then I examine every sleeping filefish I find, and all of them, including the large scrawled filefish, sometimes hold themselves in place with their mouths.

Discovering behaviors like this one is a great thrill of night diving; very seldom do the creatures of the night reef interact with divers. From the animals' point of view, the diver is a huge thing that they usually can't see because a spotlight (the diver's light) is shining directly at them. I considered the animals' perspective one night when I was visiting my octopus friend Olivia — and was inspired to shine my light on **myself** rather than on Olivia. Just as if it were a day dive, Olivia reached out an arm and grabbed my hand! She consistently responded to me at night, as long as I shined my light on myself.

I tried the same thing with a moray eel friend, but he was completely uninterested.

In fact, the only other animal I've ever interacted with on night dives was a hawksbill turtle I named Heather. Usually when a turtle is found at night, it's sleeping with its head under a ledge. If divers disturb it, it becomes frightened and swims away.

The very first night we found Heather she showed that she was different from other turtles. I'd found a green moray and was pointing it out to my buddies. Our lights disturbed the moray and its movement awakened a small turtle sleeping nearby — which, instead of swimming away, hovered around us! The little turtle had a lot of algae growing on its body and a couple of barnacles on its shell; that and its unusual behavior made me wonder if it had been kept in an aquarium.

Each of us was able to stroke this turtle on the skin around its neck, and touch its shell, and each of us was careful not to frighten it. Too soon we were low on air and had to leave. My next night dive on that site was only a couple of days later, and again, in nearly the same spot, we found the hawksbill. Again, each diver was able to touch the turtle and it stayed with us until we headed back to shore.

I'd never fed a turtle before — none had ever let me get close enough to try — but I brought a piece of fish on my next night dive and offered it to Heather. She ate it! She was fun to watch, too: if the fish was too big, she'd hold on to it with her beak and push at it with her front flippers to break it up.

For the next two months, my night dives featured interactions with

Heather with David Batalsky.

Heather the hawksbill. As time passed she lost much of her algae growth, both barnacles fell off, and she gained interest in people. We didn't need to seek her out anymore; she'd show up as soon as she saw us (except, of course, for the dive we brought video...). We could control her behavior with our lights since she'd follow a light beam: if I wanted her near me, I'd shine my light on myself; if I wanted her to swim along, I'd shine my light along the reef! And if I didn't want my fingers to get nibbled, I'd close my hand, because Heather couldn't see very well at all and once I started feeding her she tasted everything she could get her beak around, including a moray eel who was attracted by the smell of the fish. She didn't seem to like the taste of the moray, though, and she never ate hot dog, either.

I'd have loved to meet Heather in the daytime, but could only find her at night, never earlier than about nine o'clock.

I rarely can schedule late night dives, but once on a boat-based trip my buddy and I decided to try a **really** late night dive: we got into the water at 4:30 a.m. If that dive was typical, remember that just before dawn is **not** an interesting time to dive. Most of the nocturnal creatures had already retired for the night, we found, and since the sun hadn't risen by the time we left the water (just before 6 a.m.), the diurnal animals weren't active yet. I guess that this "no one active" time compensates for the "everyone active" time of dusk.

Divers who entered the water shortly after we left it, however, saw all the usual activity of the reef awakening.

THE GENTLE DIVER

When we begin diving, we're so thrilled just to be surviving under water
that we don't have the energy to think about anything else. We get comfortable in the water and begin looking around, but by then our perception
of undersea creatures (especially the sessile ones) as "furnishings" rather
than as living animals is firmly ingrained.

Exploring a coral reef differs from a terrestrial experience in four
major ways. First, we're exploring an environment three-dimensionally:
we can hover over the reef, or lower ourselves to midwater, or rest on the
bottom. Imagine being able to explore a forest with that kind of freedom!

Second, the environment we're exploring consists almost exclusive-

Olivia the octopus loved to hold my hand.

ly of animals rather than plants, even though some of the animals can't move around.

Third, almost all of these animals disregard us, in contrast to the way untamed terrestrial animals avoid people.

Fourth, the reef animals that move around have the same three-dimensional abilities we have.

But because of terrestrial experiences, and because of the appearance of the reef, many divers never perceive marine creatures as living animals that they can affect.

All I have to do is ask divers to consider how their actions are perceived by marine animals, and I can almost see the light bulbs illuminating above their heads! It's a concept that they instantly recognize as valid, but it just hadn't been in their frames of reference. Until then they'd been perfectly comfortable chasing porcupine fish to get them to puff — suddenly they realized they were deliberately eliciting a defense behavior from these fish, and that terrifying a puffer is really no great accomplishment.

But the behavior of divers can affect undersea critters in much more subtle ways. For example, predators obviously must attack a bottom-dwelling animal from above. Consequently, a diver who hovers above these animals stresses them. An octopus or moray retreats as far as possible into its den. A scorpionfish or a stingray swims away. Garden eels and jawfish sink down into their burrows as the high-off-the-bottom diver approaches.

Garden eels and jawfish are a good test of this the-higher-the-diver-the-more-stressed-the-animal theory. Find a colony of one or the other (or both), sink down to the bottom, and slowly pull yourself toward them. Calculate how far away you are when they finally retreat into their burrows. Now, hover five to ten feet off the bottom and approach them again, still

David Ritz and this Cookie Monster study each other.

Diane Widdowson and this sea snake study each other uneventfully, despite the sea snake's potent venom. Coral Sea.

slowly. Again, calculate how far you are from them when they disappear into their burrows. You'll find you can get closer if you stay low.

Next, note how graceful marine animals are. Divers with good buoyancy control who move with efficient fin motion and no sculling of their hands use the graceful movements of animals who belong in the sea, and cause no distress to their fellow reef occupants. Divers who flail about gracelessly send shock waves through the water that fish sense with their lateral lines and that they avoid as unnatural.

This conclusion is also easy to test: hover motionlessly and watch the nearby fish. Then flail around with your hands and watch the fish flee.

Another test: go to an area where fish are accustomed to being fed, and feed them. As long as you're graceful in the water they'll gather all around you; as soon as you begin to wave your hands, they'll scatter.

You'll notice that some divers and some underwater photographers have especially exciting interactions with animals and especially interesting photographs — these are the divers who are alert to the comfort of marine animals.

You may also notice that you yourself seem to have more interesting interactions with marine animals when 1) you're waiting around at the end of a dive, or 2) you haven't taken your camera, or 3) you've taken your camera but are out of film. The reason you have interactions at these times is not because marine animals like to frustrate underwater photographers, but because these are times when divers usually behave in their most mellow, least aggressive ways.

One diver told me he was waiting for his buddy to finish her film when "A cowfish swam right up to my mask! I was able to reach out and stroke him and he didn't mind at all!"

It was while she was doing a safety decompression stop that another diver looked to her side and noticed a trumpetfish lined up with the stripe on her wet suit. She couldn't believe it, so she turned horizontal in the water, and the trumpetfish turned horizontal too. She turned back vertical, and the trumpetfish stayed right with her. That trumpetfish accepted that diver as a part of the reef — because she *acted* like part of the reef.

Fish Feeding and The Gentle Diver

Most people who are interested in wild animals get a special thrill when they've persuaded those wild animals to trust them enough to take food, and divers are no exception.

The problem is that fish feeding by divers can cause situations that those divers never envisioned. For example, when we first triumphantly hand-fed the yellowtails on Bonaire, we never expected that a few years later yellowtails would mob divers every time they entered the water. . .that yellowtails would become so aggressive they effectively prevent any other fish from being fed on "their" reefs. . .that yellowtails would ruin underwater photographs, because as soon as the photographers stay in one place for a second or two (to bracket exposures, for example), the yellowtails assume they're feeding and surround the photographers, who end up with pictures edged with partial-yellowtails.

A more serious problem was caused by divers feeding eels. As

The yellowtails always know when there's a chance for a handout, but Wayne Hasson wants all the food to go to the stingrays. Grand Cayman.

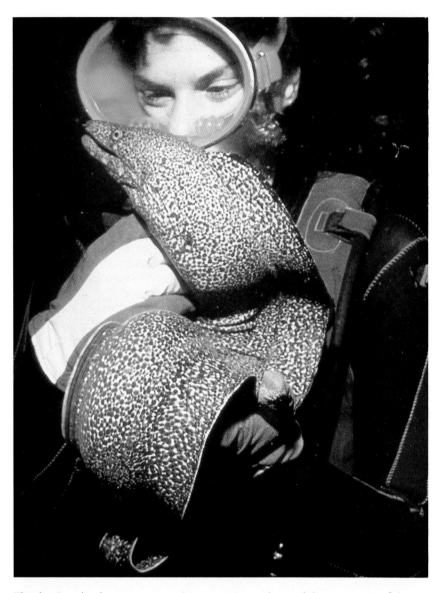

The day Spooky the moray swam into my arms and stayed there was one of the most memorable I've had, and I'm grateful Kathy was there to take the picture. Photo by Kathleen Hansell.

morays lost their "they'll attack anyone" reputation and gained their "puppy dog" reputation (neither of which, like most extremes, is valid), divers began to feed eels. Many divers used the food to lure the eel out of its den: the diver would offer food (usually fish) to the eel, then, when the eel moved toward it, the diver would move the food away. As the eel moved toward the food, the diver would move the food away. The eel would follow the food, but (since morays have no swim bladders) at a certain point it would have to begin swimming — and it would increase speed. The diver suddenly had a fast-moving eel heading straight for his or her hand. The diver would let go of the food and jerk the hand away, causing a water current that would make the food follow the hand! To make things worse, the eel, who can't see very well anyway, would follow the hand rather than the food.

People can move very quickly when they think an eel is attacking them, so most of these divers escaped unscathed — except mentally, since they wrongly perceived that the eels were attacking them.

A more serious consequence of these fiascoes is what the diver unintentionally trained the eel to do when presented with food: to go after it as quickly as possible. Add this response to the habit some divers have of wriggling their fingers to get the attention of an eel (in itself a silly thing to do, since the wriggling fingers get the attention of eel because they look like wounded prey!), and the result is eel bites.

In the last few years we've had several unpleasant diver-moray interactions on Bonaire, including one that resulted in a diver losing a finger because an eel had severed both arteries. It must be emphasized: eels are **not** vicious. Their behaviors in these cases have been **taught** to them **by divers.**

So the first guideline of Fish Feeding and The Gentle Diver is:
1. Consider what behaviors you are reinforcing!

On Little Cayman when I began to do a bit of fish feeding, the crew warned me about a grouper they'd named Tom Howard after the two people whose ears the grouper had bitten. I eyed every grouper I saw with suspicion, and was very careful to keep each one away from my head — until I saw an innocent-looking little hind eyeing me expectantly. As I concentrated on the food can, the innocent-looking little hind zoomed up, grabbed the top of my ear, and tried to pull it off of my head!!

Clearly, someone who had fed Tom Howard in the past had "taught" the grouper to grab the food and pull it out of his or her hand, not realizing that standard-eating-procedure for groupers is to suck down the whole piece of food.

In a spirit of pure unselfish philanthropy, I slugged Tom Howard. As a result of that interaction, he was then wonderfully polite to all of our group, including me, but his name was still changed to "Tom **D.** Howard."

The second guideline of Fish Feeding and The Gentle Diver is:
2. Ask for the advice of local diving professionals and take that advice.

On my first trip to Australia, I had barely set foot on the dive boat when I asked the crew for fish scraps to feed the fish. They didn't refuse me, but they "exchanged glances" and politely put me off. I waited a few days and repeated my request.

This time I got some fish scraps, and found out by experience what

The grouper Tom D. Howard was quite polite to Rod and Pam Stockton — after I gave it etiquette lessons and lunch!

happens when one brings fish into the waters of the Coral Sea: no matter how small the amount (including the little pieces I was feeding to the anemonefish to feed to the anemones), within five minutes of my putting the fish in the water a shark arrived to investigate the scent. Luckily, the sharks were all small ones (less than six feet long) and they all left when they saw divers.

Although the *Coralita's* crew obviously didn't want to give me unsolicited advice, they protected me from myself in a subtle way by giving me only small pieces of fish. Perhaps they were afraid that if I'd had bigger pieces I'd try to feed one of those sharks. Well, they may have been right . . . if I'd had Valerie Taylor's chain mail suit and gloves, and if I'd had a piece of fish two feet or so long, I might have tried to feed one of those investigating sharks.

As it was, however, I had no problems resisting my shark-feeding desires — until we discovered a ledge with four small (less than three feet long) reef whitetip sharks resting on it. Reef whitetips are docile-appearing sharks that don't have to swim to breathe. I considered the Golden Rule of Boat Diving (Don't Do Anything That Has the Foreseeable Consequence of Ending the Trip), decided that the sharks were too small to hurt me enough to break the rule, and slipped on my gloves.

The third guideline of Fish Feeding and The Gentle Diver is:
3. Always wear brightly-colored gloves if you'd like the animal you're feeding to be able to distinguish between your fingers and the food.

As soon as I opened the fish container, the little sharks caught the scent and began swimming in quick circles around the ledge. I thought, "What will I do if they all come for the food at once?" and immediately expertly eradicated that thought from my mind and held out a piece of fish. One (only one!) little shark swam right up to my hand and took the food. Success! We repeated these actions a few times, the little sharks and I, and soon ran out of large enough pieces of food. As I replaced the cover on the fish container, one of the five-footers cruised by. It can't hurt to have a little dumb luck now and then . . .

Just as fish scent in Australian waters results in the appearance of sharks, fish scent in Bonairean waters results in the appearance of morays. These effects lead us to the fourth guideline of Fish Feeding and The Gentle Diver:
4. If you must use fish to feed the fish, carry it in a **can** so you can control the scent, and so you can drop it if you want to. (Plastic containers are buoyant in seawater.)

Even a can may leak a bit of scent, and in my early days of eel-feeding, a spotted moray once swam toward me rather aggressively.
The fifth guideline of Fish Feeding and The Gentle Diver is actually a corollary of Guideline number 4:
5. Avoid wearing *Eau de Dead Fish* cologne.
I dropped the can (and the fish scent) and backed away, very glad I could separate myself from the fish.

One of the simplest ways of following guidelines #4 and #5 is to use something besides fish to feed the fish. In general I don't like bread, cheese-in-squirt-cans, or anything else that is uncontrollable or clouds the water — because it's uncontrollable and clouds the water. One of my favorite fish foods is hot dog: it's easy to obtain everywhere except on a live-aboard dive boat, it's relatively inexpensive, and it doesn't have a scent that attracts predators.

The only problem with hot dog is that it's unfamiliar to fish; unless they're already acclimated to taking food from divers, they won't respond to it immediately. A little patience will usually result in the interest of the more opportunistic feeders, such as the wrasses and yellowtails.

All the diver has to do is face down current and break off tiny pieces of the hot dog until fish begin to follow the trail. Then, holding the hot dog very steadily in front of you, wait for the fish to realize that nibbling on that food in your hand is the only way they're going to get any more of it!

Tom D. Howard's relatives, the groupers, want to see the entire piece of food so they know they can suck it down all at once. In that case the diver has to place the food close by and entice the fish to gobble it down.

Angelfish shaped like the Caribbean's French and grey angels usually respond quickly to attempts to feed them; angels shaped like the Caribbean's queen and blue angels rarely if ever come in to take food from divers.

The more knowledge a diver has about the habits of various fish, the better success that diver will have in feeding those fish and also in simply getting close to them.

One of the most enjoyable ways to use such knowledge is to attract fish without carrying food. For example, goatfish, some wrasses, and even some jacks forage on sand flats; a diver who pats up a cloud of sand is likely to attract a following of sand-feeders.

Barracudas are instinctively interested in anything shiny, white, or fluttering. If you wriggle a shiny knife, wave a white towel, or even wave your hand so that it catches the sunlight, you're likely to attract a barracuda. It is most sensible to involve yourself in barracuda-attracting when you're in water clear enough to let the 'cuda see what's attracting it **before** it gobbles down your towel (or your hand). I once settled on a sand bottom, tucked my hands away, and motionlessly watched a barracuda. The creature slowly turned toward me, then, teeth flashing in the sunlight, began to swim toward me. Frantically I wondered what I was doing to attract it — and what I could do to **stop** attracting it — when I noticed something out of the corner of my eye: my dive buddy, behind me, waving his hand.

Perhaps the sixth guideline of Fish Feeding and The Gentle Diver should be:

6. Be certain you know what your buddy is doing.

UNDERWATER PHOTOGRAPHY AND THE GENTLE DIVER

Imagine for a minute that you're a queen angelfish. You're foraging for food, keeping an eye out for queen angelfish predators, and along comes this thing about six times longer than you are. It's making horrendous noises, it's got three eyes and two of them extend in front of it on stalks — and as soon as it sees you it starts swimming right at you!! Terrified, you

Every photographer around wanted to shoot this bumphead wrasse, but we all managed to stay off the bottom anyway!

freeze for a second, and light flashes from one of the eyes! You regain your wits and swim as quickly as possible away from this monster, but it chases you, frightening you even more. Behind it is a cloud of sand, apparently excreted by the monster, and you swim frantically into the softly enveloping folds of the excretion and cower near a coral head.

The monster seems not to want to enter this cloud of its own making; it turns away. You gulp a sigh of relief and resume foraging.

Now imagine that you are a small crab, cuddled near the protecting tentacles of an anemone, munching away on some algae. Along comes a creature so enormous you can't see all of it at once. It extends one of its claws and grabs you — gently, it must be admitted — but then lifts you away from your home and drops you hundreds of your body lengths away from the security of the bottom. Images of the mouths of fierce fish scream through your mind as you make yourself as small as possible to sink more quickly. The monster extends a claw and thrusts a black machine at you. There is a flash of light. You are blinded.

When your sight returns, you are back on the bottom — but your anemone is nowhere in sight. A large fish appears, opens its mouth, and —

It seems to be a fact of diving that, no matter how kind or gentle a person is, when that person is behind an underwater camera system, he or she thinks only of the photographic results and rarely of the animal subjects.

And yet, this kind of approach is **not** the approach that results in the best photographs! It's obvious, here in print, that a frightened animal is unlikely to look its best — but the tricky part is to remember that fact during a photo outing.

When I guide divers, I concentrate on animals and the possibilities of interactions between the divers and those animals. I try to envision the interactions from the animals' points of view: what makes them comfortable, etc. It's at these times that I see most clearly how underwater photographers stress marine animals — and lose opportunities for great photographs — by being too aggressive.

So, here are a few hints for underwater photographers that result in better photographs **and** a healthier marine environment.

Most importantly, the photographer should be properly weighted. A neutrally buoyant diver gets crisp photographs because no sand is kicked up; also, since such a diver floats gracefully along the reef, the animals aren't disturbed. The photographer can then drift imperceptibly into range to get lovely portraits of fish.

When seeking fish photos, take the advice of Dave Woodward: "If you chase fish, you get pictures of fish tails." It's difficult, when we see a beautiful fish, not to zoom directly at it for a picture — but that frightens the fish. The most efficient fish-stalking method is to swim along a path not quite parallel to the subject, so that you get closer to the fish gradually. Quite often this gentle stalking results in the fish turning to look at you with curiosity rather than fear, and you're rewarded with a nice shot.

*John Pohle's camera system doesn't keep him from attaining neutral buoyancy —
and a crown of jacks.*

Whenever possible, choose your lenses so that a three to four foot
subject-to-camera distance will fill most of the frame with the subject.
With a Nikonos system, that means a subject at least three feet long for the
15mm and 20mm lenses and at least a foot long for the 28mm or 35mm
lenses. Smaller subjects generally get lost in the picture.

You could, in theory, fill a wide-angle frame with a foot-long fish if
you shot very close up; in practice, you wouldn't be likely to get close
enough to the fish.

A final suggestion for wide-angle or "normal" lens photography: shoot
subjects that don't move. Sponges, soft corals, and stony coral heads all
make lovely subject matter, especially if you have a diver or two in the picture to establish size relationships. Stationary subjects give you opportunities to experiment with composition and practice your distance estimates;
they are large so they fill the frame; they reward you with bright colors.

Close-up or macro photography is challenging in a different way.

Any time you're using a framer or wand, try to keep in mind how threatening it appears to your subjects. If the subject is attached to a coral head, or if the coral head **is** the subject, the framer can easily damage the coral. A photographer who leans a framer against a coral head is also likely to bend the framer or even dislodge the lens from the camera and cause a flood. The best situation is one in which the marine environment and the photography equipment don't make contact.

Christmas tree worms and their relatives make wonderful macro subjects. With the strobe attached to the camera bracket, slowly move the framer until it's not quite touching the base of the whorls, and shoot. This not only avoids damaging the coral, it keeps the tips of the worm's whorls in focus. (If the strobe is hand-held, place it into position first, then put the framer into position and take the picture.)

When shooting with extension tubes or the close-up kit, I concentrate on color and composition. Nudibranchs are frustrating in those respects, since they like to crawl around on dead coral, and, in their preferred positions, can only be photographed from the top. Boring! The solution is to move the subject.

Remember that anyone who takes the responsibility of moving animals should also take the responsibility of replacing them.

When posing a subject, I look for a place where the animal is comfortable and where it makes a pleasing picture that can be taken without damaging anything.

A sand bottom, or a sand-covered-with-nice-green-algae bottom, is a good location for photographing nudibranchs. The framer can dig into the sand below the nudi for a nice shot of the nudibranch looking straight at the lens, or the nudi from the side. Neither the nudi nor the sand is damaged this way, and after the photos are taken the nudibranch can be carefully replaced on whatever nonphotogenic area it was found.

This method works equally well with crabs and other critters that don't move very quickly. I've even been able to photograph small scorpionfish this way.

Some subjects can also be posed nicely on sponges, especially the tips of finger sponges.

Probably the worst place to pose any animal for macro photography is on a living coral head. Many critters, including nudibranchs, won't tolerate being posed there and will drive the photographer mad trying to keep them in position for the shot. The pose the animal ends up in is usually not very interesting, and it's almost impossible to shoot such a picture without damaging the coral head.

Another unfortunate method of posing macro critters is tossing the animals up in mid-water and fitting the framer around them as they fall. That situation, from the animal's point of view, was delineated at the beginning of this chapter. From the photographer's point of view, it creates two additional problems. First, the position is completely unnatural: arrow

I've taken the framer off the close-up kit and dug the end of the wand into the sand to photo this scorpionfish without stressing it. Photo by C. David Batalsky.

crabs extend all their legs above their bodies and other crabs bunch their legs under their bodies so they can return to the bottom most efficiently. Nudibranchs scrunch up. All these critters look to me as if they're screaming — which they probably would be, if they could.

Second, while the creature is falling through the water, it's vulnerable to being gobbled up by the nearest yellowtail or grouper, which, along with being fatal for the subject, limits the photographer's options for bracketing the shots or varying the poses.

The close-up kit is perfect for shooting portraits of smaller fish, such as butterflyfish and damselfish, but usually the fish aren't willing to swim into the framer. An easy solution to this problem is to use the close-up kit at night when the fish are sleeping and more likely to stay in place while the framer is placed around them.

Another close-up kit trick is to remove the framer and use just the wand to give you the correct subject-to-camera distance. To ensure that the subject is correctly framed, I sometimes hold the loose framer in position, note where the wand should go, remove the framer, and take the photo. This system also avoids the problem of framer shadow in the photo.

Video or movies are most interesting if the subjects are moving — yet a series of scenes of fish swimming frantically away from the camera is just as boring as a series of still shots of fish tails. Involving a buddy to feed the fish helps, as does knowing enough about marine animal behavior to recognize animal interactions. One fascinating subject for under-

Tom triggered the shutter at the perfect moment for this photo of Velvet the eel and me. Photo by Tom Downs.

water video is cleaner stations; when the cleaners and their clients are fully involved with each other they'll let the camera come very close.

Spawning behaviors also make good video subjects, and showing "Sex Under The Sea" is a great way to guarantee audience interest!

The videographer must be just as careful about neutral buoyancy as the still photographer. With video it's much more obvious to the audience who kicked up that cloud of sand, so both the aesthetics of the tape and the ego of the videographer require good buoyancy control!

Overweighted underwater photographers ruin their wet suits, the environment they shoot in, and usually their own photographs. Aggressive underwater photographers get photos of fish tails and lose invertebrate subjects as they're eaten by predators.

Gentle underwater photographers, by understanding the environment in which they're working, can get interesting, high-quality photos — and can do so again and again.

The bottom line is simple: the more alert the underwater photographer is to the safety and comfort of the marine environment, the better will be the photographic results.

Appendix 1:
SKILLS AND EQUIPMENT

Observing marine animals in their own environment requires a few skills.

At the very minimum, the observer must be able to use a mask and snorkel: the mask provides human eyes with the air space they need to focus and see properly; the snorkel allows the observer to watch continuously without surfacing for air.

Fins are necessary for efficient propulsion in the water. When the observer kicks below the surface of the water, the animals are much less disturbed than they are by kicks that break the water's surface.

There are times when surface snorkeling is the best way to watch reef animals. When Dr. Eugenie Clark and her assistants observe the behaviors of razorfish, for example, they often see more of the fishes' activity with mask and snorkel than they could with scuba.

Still, the animals accessible to the snorkeler's eyes are limited to those that inhabit the first twenty feet or so of depth, in waters with good visibility, in areas where the surface is calm.

An observer using scuba can extend his or her underwater range considerably — but, of course, even scuba has its limitations. Anyone using scuba equipment should be certified, properly trained, and properly equipped.

Watching and interacting with marine animals doesn't require a great deal of movement, so even divers who don't normally chill under water may feel cold. The only time I can remember not getting cold under water was on the Bahamas Bank — a huge area where the maximum depth is about twenty feet — in August.

At all other times I wear a wet suit. On particularly long dives, I add (to my usual 3/16" full suit) a hooded 1/8" vest. This combination keeps me comfortable under almost all tropical circumstances — due not only to the additional thickness of neoprene, but because my head and neck, areas particularly susceptible to cold, are covered.

Understanding how to use scuba equipment is more than simply knowing how to breathe under water. **The most critical skill in diving is good buoyancy control.** Divers who are too heavy or too light in the water can't stop to look at anything — they have to kick to avoid sinking to the bottom or floating to the surface. This problem should be obvious, but very often it's not; many divers never realize how hard they're struggling to stay in place. Sometimes, though, they realize they're not seeing as many interesting things as the other people on the dive.

Here are some hints for comfortable diving:

Proper weighting is essential. It's true that if you're overweighted, you can compensate with your buoyancy control device. **But** for every pound you're overweighted, you have to carry extra air in your BCD for the entire dive, **and** the amount of extra air varies with depth. **So,** if you're carrying unnecessary weight, you have to add **more** air when you get to depth, and as you ascend along the reef slope, the air in your BCD expands more, and you have more trouble getting neutral at each new depth.

Also, the act of putting air into your BCD with some types of power inflators affects marine animals. The newer power inflators have a "soft" sound, but the older ones have a percussive sound that is much like a grouper's threat noise (the grouper does it with its swim bladder). When I used my loud inflator on San Salvador's reefs, all the groupers would scatter momentarily. The goal is to have a quiet power inflator, to use it only when necessary, and for it to be necessary as seldom as possible.

Try this: at the end of a dive, when you have as little air left in your tank as you usually do, go to about ten feet of depth (or as shallow as you want to be comfortable). Make sure all the air is out of your BCD, hover horizontally not touching the bottom, and **stop kicking.**

If you sink on a full inhalation, you're overweighted. Take a pound or two off your weight belt and try the test again. Repeat this process until you hover in place or sink very gently.

If you bob to the surface, add a pound or two to your weight belt and repeat the process until you hover in place or sink very gently.

Quite often just being properly weighted adds significantly to a diver's bottom time and decreases after-dive fatigue. Struggling against improper weighting is a lot of work!

When you're satisfied that you're using the correct amount of weight, **log the information:** how much weight, fresh or salt water, type and thickness of wetsuit, aluminum or steel tank. Once you have your weight information for one set of circumstances, it's easy to extrapolate for other conditions.

Being perfectly weighted doesn't mean you don't need to use your BCD. Since compressed air has weight, tanks lose weight as they lose air. Whatever type of tank you use, you must be negative at the beginning of the dive to be properly weighted at the end of the dive. You'll have to put some air into your BCD at first, and, as your tank becomes more buoyant, let air out of your BCD to compensate.

A wetsuit provides additional complications. It insulates with air bubbles, so the deeper it goes, the more it compresses, resulting in less buoyancy. If you're wearing a wetsuit and diving along a drop-off, you need even more air in the BCD at the beginning of the dive, when you're deeper, and less air at the end, when you're shallower.

You can tell if your buddies are using the right amount of air in their BCDs by watching their attitude in the water. If they swim in a head-up position, they probably need to add some air; if they're head-down, they need to dump some air. To test yourself, simply stop kicking. If you begin to rise (usually feet-first), it's time to dump air (remember that you probably won't be **able** to dump the air unless you return to a head-up position); if you sink quickly, add some air. There's no reason that leg power should need to replace proper buoyancy control.

The best way to hone your buoyancy skills is to **practice.** Practice in a pool between dive trips. Practice hovering vertically, hovering horizontally, hovering in a sitting position, without touching the bottom or breaking the surface. Practice in your wet suit for the added difficulty, even if the water is warm. Practice with too much weight so you can feel the additional awkwardness, and with correct weighting (but remember that, unless it's a salt-water pool, you'll need more weight when diving in the sea). Practice with your hands clasped behind your back, or while holding a light or camera. Practice until you can sense when you're sinking or rising. Practice until you're totally bored.

The results will be worth it.

Appendix 2:
BOOKS OF INTEREST

GENERAL

Coulombe, Deborah A. *The Seaside Naturalist, a Guide to Nature Study at the Seashore.* Englewood Cliffs, NJ 07632: Prentice-Hall, Inc., 1984. Good basic information, primarily about animals found along U.S. coastlines. Black-and-white drawings.

Emery, Alan. *The Coral Reef.* The Canadian Broadcasting Corporation, 1981. Published in conjunction with the CBC television program of the same name. A good introduction to the coral reef with good photos well-reproduced on high-quality paper.

Kaplan, Eugene H. *A Field Guide to Coral Reefs of the Caribbean and Florida* (of the Peterson Field Guide series). Boston: Houghton-Mifflin Company, 1982. Every time I read this book I learn something. It's an introduction to the animals of the areas covered, so it doesn't even attempt to identify them all, but the information about interrelationships and behaviors is great. A few color plates, some black-and-white photos and black-and-white illustrations.

Niessen, Thomas M. *The Marine Biology Coloring Book.* New York: Barnes and Noble (Harper and Row), 1982. When I first saw this book I thought it was a for kids — I was wrong, though I have as much fun as a kid when I read it! It's divided by animal groups and subdivided by behaviors, and the coloring is to help the reader recognize "similarity and diversity of form and function in marine organisms."

Nybakken, James W. *Marine Biology: An Ecological Approach.* New York: Harper and Row, 1982. Recommended to me by Dr. Eugenie Clark (she uses it as a textbook in some of her classes), this book is more technical than most of the others on this list. It's organized around various ecosystems (plankton, the deep sea, tropical communities, symbiotic relationships) rather than phylum-by-phylum, and is interesting — but not light — reading.

Scarr, Dee. *Touch the Sea*. PADI, 1987. Naturally I hope you're already familiar with this book, but if you're not I strongly recommend it. It's my mother's favorite book.

INVERTEBRATES

Meinkoth, Norman A. *The Audubon Field Guide to North American Seashore Creatures*. New York: Alfred A. Knopf, 1981. This has little information about behaviors, and is most useful for those wanting to identify invertebrates of the temperate waters around the U.S. The paper on which the text is printed is so thin that I constantly fear I'll wear it out by reading it, but the photos are in color and on sturdy paper.

Colin, Dr. Patrick I. *Caribbean Reef Invertebrates and Plants*. T. F. H. Publications, 1978. This book was out of print until 1988, when T.F.H. reprinted it. Thanks, T.F.H.!! *Caribbean Reef Invertebrates and Plants* is the most comprehensive invertebrate book I know of that covers the Caribbean area. It has general information on the groups of animals and specifics about every critter shown in a photograph, and it has lots of color photos (taken in the sea, too, not in aquaria). I was frustrated by a few captions that read "Unidentified . . ." until I realized that those critters weren't identified anywhere else, either. The binding is high-quality and the paper and reproduction is good.

Sefton, Nancy, and Steven K. Webster. *Caribbean Reef Invertebrates*. A special publication of the Monterey Bay Aquarium Foundation, 1986. The authors have filled the gap in Caribbean invertebrate identification sources with a book of high-quality photos taken in the sea and high-quality information about the critters in the photos.

Voss, Gilbert L. *Seashore Life of Florida and the Caribbean*. Miami: E. A. Seemann Publishing, Inc., 1976. This book presents good information. It's not great for identification, though, since there are only a few color plates and the rest of the illustrations are line drawings. Available in both hard and soft cover.

Zeiller, Warren. *Tropical Marine Invertebrates of Southern Florida and the Bahama Islands*. New York: John Wiley and Sons, 1974. Since Zeiller was curator of the Miami Seaquarium when he published this book, it's not surprising that most of the photographs are of animals in aquaria, and that what information he gives about the habits of the critters relates to their lives in captivity. Still, the photos display the animals well and the paper is high quality.

FISH

Randall, John E. *Caribbean Reef Fishes*. New Jersey: T. F. H. Publications, 1968. Randall deliberately left out the gobies and blennies and a few other groups of fish divers see frequently, but he covers the groups he does cover accurately and comprehensively. There is a photograph of each fish in the text and many of them are in color. The discussions are mostly technical and geared to the identification of a dead fish, but some of the information is useful even for the casual sport diver.

Stokes, F. Joseph. *Divers and Snorkelers Guide to the FISHES and Sea Life of the Caribbean, Florida, Bahamas, and Bermuda*. The Academy of Natural Sciences of Philadelphia, 1984. I love this book. It's small, it's hardbound and well-bound (I've used my three copies so much they should be falling apart, and they're holding together fine), and there is much more information presented than is at first apparent. This book is the only true field guide to Caribbean fish in that it has a special section on "Coping with a 'mystery' fish" (with which I was able to instantly identify my own personal mystery fish). The bulk of the book consists of fish drawings accompanied by simple identifying text. Hidden in the back are various notes about groups of fish and behaviors and physiological information, and even a glossary — all presented understandably! This book even has a mini-index on the inside back cover for quick reference. I wish such a book could be created for fish and invertebrates of all major areas!

Thresher, Ronald E. *Reef Fish: Behavior and Ecology on the Reef and in the Aquarium*. St. Petersburg, Florida: Palmetto Publishing Company, 1980. Years ago I overheard someone ask Bonairean biologist Roberto Hensen if there was a book that discussed behaviors of Caribbean fish. "Too bad there isn't," I thought to myself, then heard Robbie recommending *Reef Fish*. Although Thresher limits himself to discussing fish that are commonly kept in aquaria, the information he presents about those fish is fascinating.